UGLY'S™

ELECTRICAL REFERENCES

by

GEORGE V. HART

*

REVISED 1981, 1984, 1987, 1990, 1993, 1996, 1999, 2002

*

PRINTED IN U.S.A.

* * *

DISTRIBUTED BY:
Burleson Distributing Corp. • 3501 Oak Forest Drive • Houston, Texas 77018
(713) 956-6666 • (800) 531-1660 • Fax (713) 956-6576
E-mail: uglys@uglyselectrical.com
Website: www.uglyselectrical.com

TABLE OF CONTENTS

(continued next page)

TABLE OF CONTENTS (continued)

(continued next page)

TABLE OF CONTENTS (continued)

OHM'S LAW

The rate of the flow of the current is equal to electromotive force divided by resistance.

I = Intensity of Current = Amperes
E = Electromotive Force = Volts
R = Resistance = Ohms
P = Power = Watts

The three basic Ohm's law formulas are:

$$I = \frac{E}{R} \qquad R = \frac{E}{I} \qquad E = I \times R$$

Below is a chart containing the formulas related to Ohm's law.
To use the chart, from the center circle, select the value you need to find, I (Amps), R (Ohms), E (Volts) or P (Watts). Then select the formula containing the values you know from the corresponding chart quadrant.

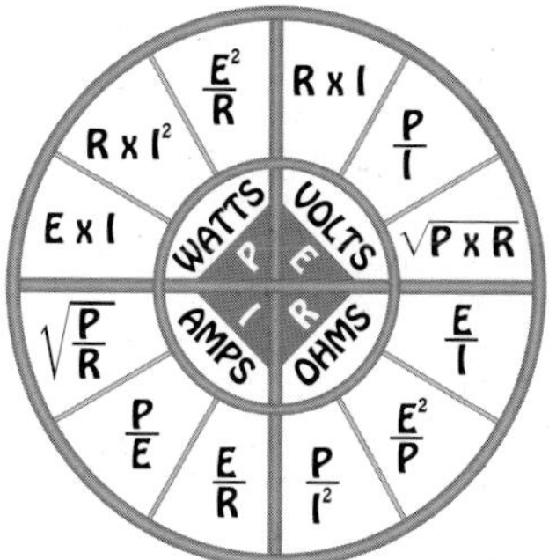

Example:
An electric appliance is rated at 1200 Watts, and is connected to 120 Volts. How much current will it draw?

$$\textbf{Amperes} = \frac{\textbf{Watts}}{\textbf{Volts}} \qquad I = \frac{P}{E} \qquad I = \frac{1200}{120} = 10\text{ A}$$

What is the Resistance of the same appliance?

$$\textbf{Ohms} = \frac{\textbf{Volts}}{\textbf{Amperes}} \qquad R = \frac{E}{I} \qquad R = \frac{120}{10} = 12\ \Omega$$

OHM'S LAW

In the preceding example, we know the following values:

I = amps = 10 R = ohms = 12Ω

E = volts = 120 P = watts = 1200

We can now see how the twelve formulas in the Ohm's Law chart can be applied.

$$\text{AMPS} = \sqrt{\frac{\text{WATTS}}{\text{OHMS}}} \qquad I = \sqrt{\frac{P}{R}} = \sqrt{\frac{1200}{12}} = \sqrt{100} = 10\text{A}$$

$$\text{AMPS} = \frac{\text{WATTS}}{\text{VOLTS}} \qquad I = \frac{P}{E} = \frac{1200}{120} = 10\text{A}$$

$$\text{AMPS} = \frac{\text{VOLTS}}{\text{OHMS}} \qquad I = \frac{E}{R} = \frac{120}{12} = 10\text{A}$$

$$\text{WATTS} = \frac{\text{VOLTS}^2}{\text{OHMS}} \qquad P = \frac{E^2}{R} = \frac{120^2}{12} = \frac{14{,}400}{12} = 1200\text{W}$$

$$\text{WATTS} = \text{VOLTS} \times \text{AMPS} \qquad P = E \times I = 120 \times 10 = 1200\text{W}$$

$$\text{WATTS} = \text{AMPS}^2 \times \text{OHMS} \qquad P = I^2 \times R = 100 \times 12 = 1200\text{W}$$

$$\text{VOLTS} = \sqrt{\text{WATTS} \times \text{OHMS}} \qquad E = \sqrt{P \times R} = \sqrt{1200 \times 12} = \sqrt{14{,}400} = 120\text{V}$$

$$\text{VOLTS} = \text{AMPS} \times \text{OHMS} \qquad E = I \times R = 10 \times 12 = 120\text{V}$$

$$\text{VOLTS} = \frac{\text{WATTS}}{\text{AMPS}} \qquad E = \frac{P}{I} = \frac{1200}{10} = 120\text{V}$$

$$\text{OHMS} = \frac{\text{VOLTS}^2}{\text{WATTS}} \qquad R = \frac{E^2}{P} = \frac{120^2}{1{,}200} = \frac{14{,}400}{1{,}200} = 12\Omega$$

$$\text{OHMS} = \frac{\text{WATTS}}{\text{AMPS}^2} \qquad R = \frac{P}{I^2} = \frac{1200}{100} = 12\Omega$$

$$\text{OHMS} = \frac{\text{VOLTS}}{\text{AMPS}} \qquad R = \frac{E}{I} = \frac{120}{10} = 12\Omega$$

SERIES CIRCUITS

A SERIES CIRCUIT is a circuit that has only one path through which the electrons may flow.

RULE 1: **The total current in a series circuit is equal to the current in any other part of the circuit.**

TOTAL CURRENT $I_T = I_1 = I_2 = I_3$, **etc.**

RULE 2: **The total voltage in a series circuit is equal to the sum of the voltages across all parts of the circuit.**

TOTAL VOLTAGE $E_T = E_1 + E_2 + E_3$, **etc.**

RULE 3: **The total resistance of a series circuit is equal to the sum of the resistances of all the parts of the circuit**

TOTAL RESISTANCE $R_T = R_1 + R_2 + R_3$, **etc.**

FORMULAS FROM OHM'S LAW

$$\text{AMPERES} = \frac{\text{VOLTS}}{\text{RESISTANCE}} \quad \text{OR} \quad I = \frac{E}{R}$$

$$\text{RESISTANCE} = \frac{\text{VOLTS}}{\text{AMPERES}} \quad \text{OR} \quad R = \frac{E}{I}$$

$$\text{VOLTS} = \text{AMPERES} \times \text{RESISTANCE} \quad \text{OR} \quad E = I \times R$$

EXAMPLE: Find the total voltage, total current, and total resistance of the following series circuit.

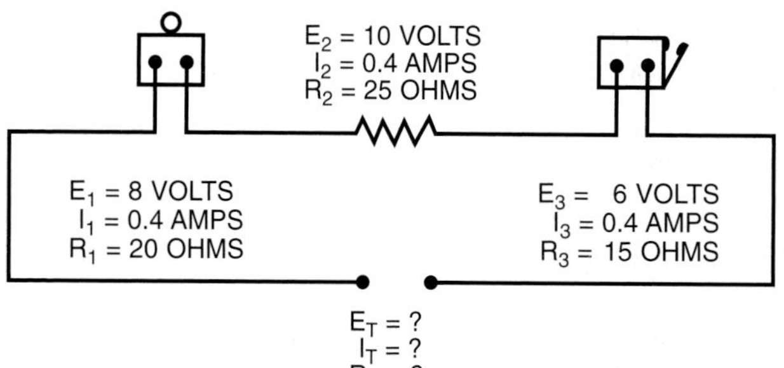

$E_T = ?$
$I_T = ?$
$R_T = ?$

(continued next page)

SERIES CIRCUITS

$$E_T = E_1 + E_2 + E_3$$
$$= 8 + 10 + 6$$
$$E_T = 24 \text{ VOLTS}$$

$$I_T = I_1 = I_2 = I_3$$
$$= 0.4 = 0.4 = 0.4$$
$$I_T = 0.4 \text{ AMPS}$$

$$R_T = R_1 + R_2 + R_3$$
$$= 20 + 25 + 15$$
$$R_T = 60 \text{ OHMS}$$

EXAMPLE: Find E_T, E_1, E_3, I_T, I_1, I_2, I_4, R_T, R_2, AND R_4. Remember that the total current in a series circuit is equal to the current in any other part of the circuit.

$E_1 = ?$
$I_1 = ?$
$R_1 = 72$ OHMS

$E_3 = ?$
$I_3 = 0.5$ AMPS
$R_3 = 48$ OHMS

$E_2 = 12$ VOLTS
$I_2 = ?$
$R_2 = ?$

$E_4 = 48$ VOLTS
$I_4 = ?$
$R_4 = ?$

$E_T = ?$ $I_T = ?$ $R_T = ?$

$$I_T = I_1 = I_2 = I_3 = I_4$$
$$I_T = I_1 = I_2 = 0.5 = I_4$$
$$0.5 = 0.5 = 0.5 = 0.5 = 0.5$$
$I_T = 0.5$ AMPS $I_2 = 0.5$ AMPS
$I_1 = 0.5$ AMPS $I_4 = 0.5$ AMPS

$$E_T = E_1 + E_2 + E_3 + E_4$$
$$= 36 + 12 + 24 + 48$$
$$E_T = 120 \text{ VOLTS}$$

$$R_T = R_1 + R_2 + R_3 + R_4$$
$$= 72 + 24 + 48 + 96$$
$$R_T = 240 \text{ OHMS}$$
$$R_2 = \frac{E_2}{I_2} = \frac{12}{0.5}$$
$$R_2 = 24 \text{ OHMS}$$

$$E_1 = I_1 \times R_1$$
$$= 0.5 \times 72$$
$$E_1 = 36 \text{ VOLTS}$$

$$E_3 = I_3 \times R_3$$
$$= 0.5 \times 48$$
$$E_3 = 24 \text{ VOLTS}$$

$$R_4 = \frac{E_4}{I_4} = \frac{48}{0.5}$$
$$R_4 = 96 \text{ OHMS}$$

PARALLEL CIRCUITS

A PARALLEL CIRCUIT is a circuit that has more than one path through which the electrons may flow.

RULE 1: The total current in a parallel circuit is equal to the sum of the currents in all the branches of the circuit.

TOTAL CURRENT $I_T = I_1 + I_2 + I_3$, etc.

RULE 2: The total voltage across any branch in parallel is equal to the voltage across any other branch and is also equal to the total voltage.

TOTAL VOLTAGE $E_T = E_1 = E_2 = E_3$, etc.

RULE 3: The total resistance of a parallel circuit is found by applying OHM'S LAW to the total values of the circuit.

$$\text{TOTAL RESISTANCE} = \frac{\text{TOTAL VOLTAGE}}{\text{TOTAL AMPERES}} \quad \text{OR} \quad R_T = \frac{E_T}{I_T}$$

Example: Find the total current, total voltage, and total resistance of the following parallel circuit.

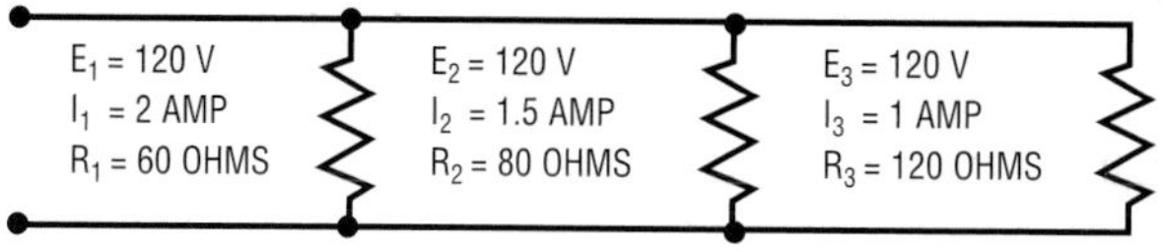

$E_1 = 120$ V, $I_1 = 2$ AMP, $R_1 = 60$ OHMS

$E_2 = 120$ V, $I_2 = 1.5$ AMP, $R_2 = 80$ OHMS

$E_3 = 120$ V, $I_3 = 1$ AMP, $R_3 = 120$ OHMS

$$I_T = I_1 + I_2 + I_3 = 2 + 1.5 + 1$$
$$I_T = 4.5 \text{ AMPS}$$

$$E_T = E_1 = E_2 = E_3 = 120 = 120 = 120$$
$$E_T = 120 \text{ VOLTS}$$

$$R_T = \frac{E_T}{I_T} = \frac{120 \text{ VOLTS}}{4.5 \text{ AMPS}} = 26.66 \text{ OHMS RESISTANCE}$$

NOTE: In a parallel circuit, the total resistance is always less than the resistance of any branch. If the branches of a parallel circuit have the same resistance, then each will draw the same current. If the branches of a parallel circuit have different resistances, then each will draw a different current. In either series or parallel circuits, the larger the resistance, the smaller the current drawn.

PARALLEL CIRCUITS

To determine the total resistance in a parallel circuit when the total current and total voltage are unknown:

$$\frac{1}{\text{TOTAL RESISTANCE}} = \frac{1}{R_1} + \frac{1}{R_2} + \frac{1}{R_3} \quad \text{AND ETC.}$$

EXAMPLE: Find the total resistance of the following circuit:

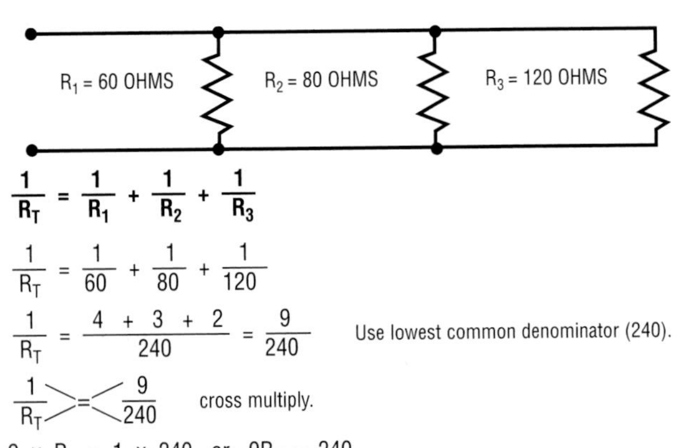

$$\frac{1}{R_T} = \frac{1}{R_1} + \frac{1}{R_2} + \frac{1}{R_3}$$

$$\frac{1}{R_T} = \frac{1}{60} + \frac{1}{80} + \frac{1}{120}$$

$$\frac{1}{R_T} = \frac{4 + 3 + 2}{240} = \frac{9}{240}$$ Use lowest common denominator (240).

$$\frac{1}{R_T} = \frac{9}{240}$$ cross multiply.

$9 \times R_T = 1 \times 240$ or $9R_T = 240$

divide both sides of the equation by 9

$R_T = 26.66$ OHMS RESISTANCE

NOTE: The total resistance of a number of EQUAL resistors in parallel is equal to the resistance of one resistor divided by the number of resistors.

$$\text{TOTAL RESISTANCE} = \frac{\text{RESISTANCE OF ONE RESISTOR}}{\text{NUMBER OF RESISTORS IN CIRCUIT}}$$

(continued next page)

PARALLEL CIRCUITS

FORMULA: $$R_T = \frac{R}{N}$$

EXAMPLE: Find the total resistance

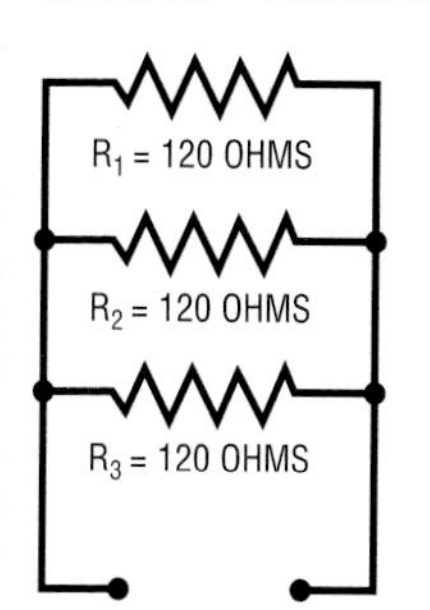

There are three resistors in parallel. Each has a value of 120 OHMS resistance. According to the formula, if we divide the resistance of any one of the resistors by three, we will obtain the total resistance of the circuit.

$$R_T = \frac{R}{N} \quad \text{OR} \quad R_T = \frac{120}{3}$$

TOTAL RESISTANCE = 40 OHMS.

NOTE: To find the total resistance of only two resistors in parallel, multiply the resistances, and then divide the product by the sum of the resistors.

FORMULA: **TOTAL RESISTANCE** $$= \frac{R_1 \times R_2}{R_1 + R_2}$$

EXAMPLE:

R_1 = 40 OHMS

R_2 = 80 OHMS

$$R_T = \frac{R_1 \times R_2}{R_1 + R_2}$$

$$= \frac{40 \times 80}{40 + 80}$$

$$R_T = \frac{3200}{120} = 26.66 \text{ OHMS}$$

COMBINATION CIRCUITS

In combination circuits, we combine series circuits with parallel circuits. Combination circuits make it possible to obtain the different voltages of series circuits, and the different currents of parallel circuits.

EXAMPLE 1. PARALLEL-SERIES CIRCUIT:
Solve for all missing values.

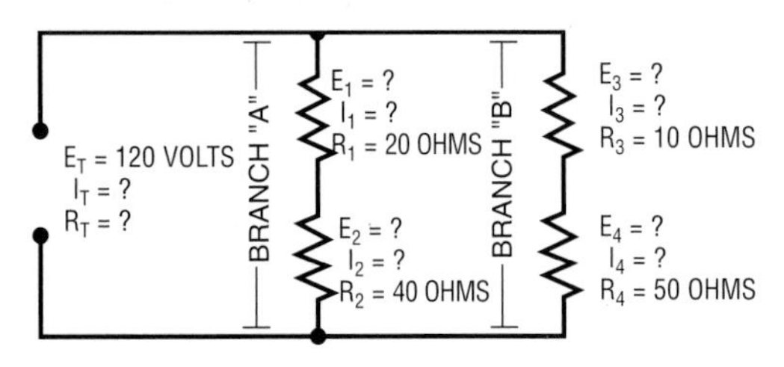

TO SOLVE:

1. Find the total resistance of each branch. Both branches are simple series circuits, so

   $\mathbf{R_1 + R_2 = R_A}$
   20 + 40 = 60 OHMS total resistance of branch "A"

   $\mathbf{R_3 + R_4 = R_B}$
   10 + 50 = 60 OHMS total resistance of branch "B"

2. Re-draw the circuit, combining resistors ($R_1 + R_2$) and ($R_3 + R_4$) so that each branch will have only one resistor.

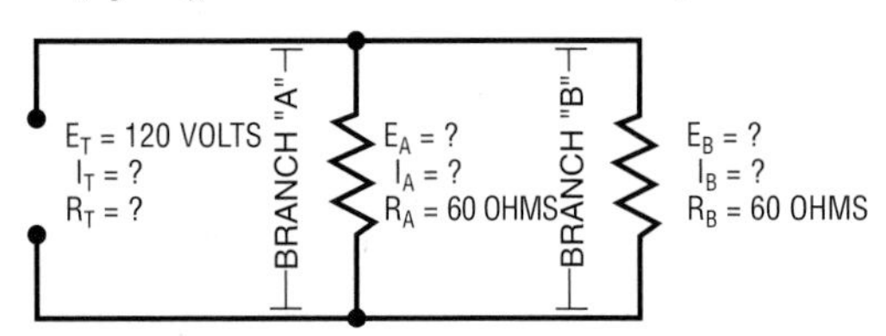

(continued next page)

COMBINATION CIRCUITS

NOTE: We now have a simple parallel circuit, so

$$\mathbf{E_T} = \mathbf{E_A} = \mathbf{E_B}$$
$$120\text{ V} = 120\text{ V} = 120\text{ V}$$

We now have a parallel circuit with only two resistors, and they are of equal value. We have a choice of three different formulas that can be used to solve for the total resistance of the circuit.

(1) $$\mathbf{R_T} = \frac{\mathbf{R_A \times R_B}}{\mathbf{R_A + R_B}} = \frac{60 \times 60}{60 + 60} = \frac{3600}{120} = 30 \text{ OHMS}$$

(2) When the resistors of a parallel circuit are of equal value,

$$\mathbf{R_T} = \frac{\mathbf{R}}{\mathbf{N}} = \frac{60}{2} = 30 \text{ OHMS} \qquad \text{OR}$$

(3) $$\frac{\mathbf{1}}{\mathbf{R_T}} = \frac{\mathbf{1}}{\mathbf{R_A}} + \frac{\mathbf{1}}{\mathbf{R_B}} = \frac{1}{60} + \frac{1}{60} = \frac{2}{60} = \frac{1}{30}$$

$$\frac{1}{R_T} = \frac{1}{30} \text{ OR } 1 \times R_T = 1 \times 30 \text{ OR } R_T = 30 \text{ OHMS}$$

3. We know the values of E_T, R_T, E_A, R_A, E_B, R_B, R_1, R_2, R_3, and R_4. Next we will solve for I_T, I_A, I_B, I_1, I_2, I_3, and I_4.

$$I_T = \frac{E_T}{R_T} \quad \text{OR} \quad \frac{120}{30} = 4 \qquad I_T = 4 \text{ AMPS}$$

$$I_A = \frac{E_A}{R_A} \quad \text{OR} \quad \frac{120}{60} = 2 \qquad I_A = 2 \text{ AMPS}$$

$$I_A = I_1 = I_2 \quad \text{OR} \quad 2 = 2 = 2 \qquad I_1 = 2 \text{ AMPS}$$

$$I_B = \frac{E_B}{R_B} = \quad \text{OR} \quad \frac{120}{60} = 2 \qquad I_B = 2 \text{ AMPS}$$

$$I_B = I_3 = I_4 \quad \text{OR} \quad 2 = 2 = 2 \qquad I_3 = 2 \text{ AMPS}$$
$$I_4 = 2 \text{ AMPS}$$

(continued next page)

COMBINATION CIRCUITS

4. We know that resistors #1 and #2 of branch "A" are in series. We know too that resistors #3 and #4 of branch "B" are in series. We have determined that the total current of branch "A" is 2 AMPS, and the total current of branch "B" is 2 AMPS. By using the series formula, we can solve for the current of each branch.

BRANCH "A"	BRANCH "B"
$\mathbf{I_A = I_1 = I_2}$	$\mathbf{I_B = I_3 = I_4}$
$2 = 2 = 2$	$2 = 2 = 2$
I_1 = 2 AMPS	I_3 = 2 AMPS
I_2 = 2 AMPS	I_4 = 2 AMPS

5. We were given the resistance values of all resistors. R_1 = 20 OHMS, R_2 = 40 OHMS, R_3 = 10 OHMS, and R_4 = 50 OHMS. By using OHM'S Law, we can determine the voltage drop across each resistor.

$\mathbf{E_1 = R_1 \times I_1}$
$= 20 \times 2$
E_1 = 40 VOLTS

$\mathbf{E_3 = R_3 \times I_3}$
$= 10 \times 2$
E_3 = 20 VOLTS

$\mathbf{E_2 = R_2 \times I_2}$
$= 40 \times 2$
E_2 = 80 VOLTS

$\mathbf{E_4 = R_4 \times I_4}$
$= 50 \times 2$
E_4 = 100 VOLTS

EXAMPLE 2: SERIES PARALLEL CIRCUIT.
Solve for all missing values.

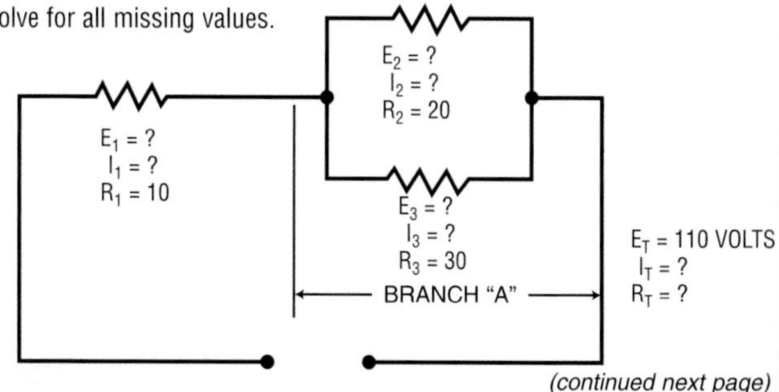

(continued next page)

COMBINATION CIRCUITS

To solve:

1. We can see that resistors #2 and #3 are in parallel, and combined they are branch "A". When there are only two resistors, we use the following formula:

$$R_A = \frac{R_2 \times R_3}{R_2 + R_3} \text{ OR } \frac{20 \times 30}{20 + 30} \text{ OR } \frac{600}{50} \text{ OR 12 OHMS}$$

2. We can now re-draw our circuit as a simple series circuit.

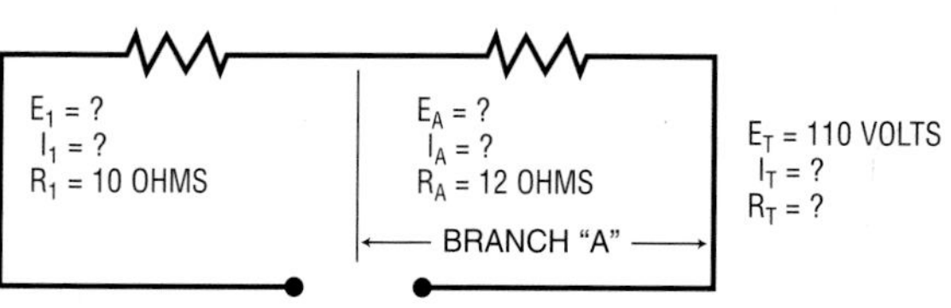

3. In a series circuit,
   $R_T = R_1 + R_A$ OR $R_T = 10 + 12$ OR 22 OHMS
   By using OHM'S Law,

   $$I_T = \frac{E_T}{R_T} = \frac{110}{22} = 5 \text{ AMPS}$$

   In a series circuit,
   $I_T = I_1 = I_A$ or $I_T = 5$ AMPS, $I_1 = 5$ AMPS, and $I_A = 5$ AMPS

   By using OHM'S Law,
   $E_1 = I_1 \times R_1 = 5 \times 10 = 50$ VOLTS
   $E_T - E_1 = E_A$ or $110 - 50 = 60$ VOLTS $= E_A$

   In a parallel circuit,
   $E_A = E_2 = E_3$ or $E_A = 60$ VOLTS
   $E_2 = 60$ VOLTS, and $E_3 = 60$ VOLTS

   By using OHM'S Law,

   $$I_2 = \frac{E_2}{R_2} = \frac{60}{20} = 3 \text{ AMPS}$$

   $$I_3 = \frac{E_3}{R_3} = \frac{60}{30} = 2 \text{ AMPS}$$

(continued next page)

COMBINATION CIRCUITS

PROBLEM:

Solve for total resistance.
Re-draw circuit as many times as necessary.
Correct answer is 100 OHMS.

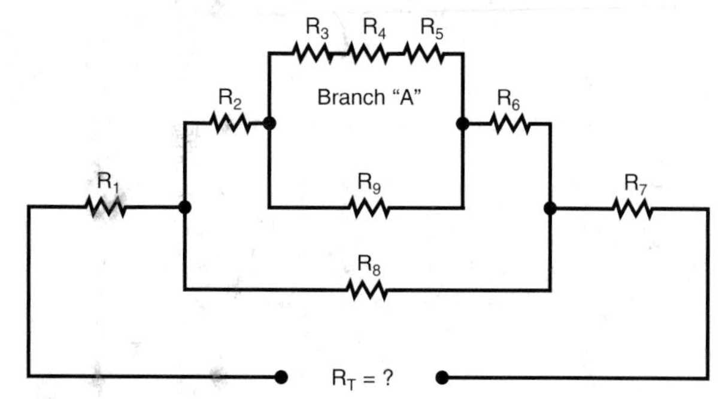

GIVEN VALUES:

R_1 = 15 OHMS				R_6 = 25 OHMS
R_2 = 35 OHMS				R_7 = 10 OHMS
R_3 = 50 OHMS				R_8 = 300 OHMS
R_4 = 40 OHMS				R_9 = 60 OHMS
R_5 = 30 OHMS				

ELECTRICAL FORMULAS FOR CALCULATING AMPERES, HORSEPOWER, KILOWATTS AND KVA

TO FIND	DIRECT CURRENT	ALTERNATING CURRENT		
		SINGLE PHASE	TWO PHASE-FOUR WIRE	THREE PHASE
AMPERES WHEN "HP" IS KNOWN	$\frac{HP \times 746}{E \times \%EFF}$	$\frac{HP \times 746}{E \times \%EFF \times PF}$	$\frac{HP \times 746}{E \times \%EFF \times PF \times 2}$	$\frac{HP \times 746}{E \times \%EFF \times PF \times 1.73}$
AMPERES WHEN "KW" IS KNOWN	$\frac{KW \times 1000}{E}$	$\frac{KW \times 1000}{E \times PF}$	$\frac{KW \times 1000}{E \times PF \times 2}$	$\frac{KW \times 1000}{E \times PF \times 1.73}$
AMPERES WHEN "KVA" IS KNOWN		$\frac{KVA \times 1000}{E}$	$\frac{KVA \times 1000}{E \times 2}$	$\frac{KVA \times 1000}{E \times 1.73}$
KILOWATTS	$\frac{E \times I}{1000}$	$\frac{E \times I \times PF}{1000}$	$\frac{E \times I \times PF \times 2}{1000}$	$\frac{E \times I \times PF \times 1.73}{1000}$
KILOVOLT-AMPERES "KVA"		$\frac{E \times I}{1000}$	$\frac{E \times I \times 2}{1000}$	$\frac{E \times I \times 1.73}{1000}$
HORSEPOWER	$\frac{E \times I \times \%EFF}{746}$	$\frac{E \times I \times \%EFF \times PF}{746}$	$\frac{E \times I \times \%EFF \times PF \times 2}{746}$	$\frac{E \times I \times \%EFF \times PF \times 1.73}{746}$

PERCENT EFFICIENCY = % EFF = $\frac{\text{OUTPUT (WATTS)}}{\text{INPUT (WATTS)}}$ POWER FACTOR = PF = $\frac{\text{POWER USED (WATTS)}}{\text{APPARENT POWER}} = \frac{KW}{KVA}$

NOTE: DIRECT CURRENT FORMULAS DO NOT USE (PF, 2, OR 1.73)
SINGLE PHASE FORMULAS DO NOT USE (2 OR 1.73)
TWO PHASE - FOUR WIRE FORMULAS DO NOT USE (1.73)
THREE PHASE FORMULAS DO NOT USE (2)

E = VOLTS
I = AMPERES
W = WATTS

TO FIND AMPERES

DIRECT CURRENT:

A. When *HORSEPOWER* is known:

$$\text{AMPERES} = \frac{\text{HORSEPOWER} \times 746}{\text{VOLTS} \times \text{EFFICIENCY}} \quad \text{or} \quad I = \frac{\text{HP} \times 746}{E \times \%\text{EFF}}$$

What current will a travel-trailer toilet draw when equipped with a 12 volt, 1/8 HP motor, having a 96% efficiency rating?

$$I = \frac{\text{HP} \times 746}{E \times \%\text{EFF}} = \frac{746 \times 1/8}{12 \times 0.96} = \frac{93.25}{11.52} = 8.09 \text{ AMPS}$$

B. When *KILOWATTS* are known:

$$\text{AMPERES} = \frac{\text{KILOWATTS} \times 1000}{\text{VOLTS}} \quad \text{or} \quad I = \frac{\text{KW} \times 1000}{E}$$

A 75 KW, 240 Volt, direct current generator is used to power a variable-speed conveyor belt at a rock crushing plant. Determine the current.

$$I = \frac{\text{KW} \times 1000}{E} = \frac{75 \times 1000}{240} = 312.5 \text{ AMPS}$$

SINGLE PHASE:

A. When *WATTS, VOLTS, AND POWER FACTOR* are known:

$$\text{AMPERES} = \frac{\text{WATTS}}{\text{VOLTS} \times \text{POWER FACTOR}} \quad \text{or} \quad \frac{P}{E \times PF}$$

Determine the current when a circuit has a 1500 watt load, a power-factor of 86%, and operates from a single-phase 230 volt source.

$$I = \frac{1500}{230 \times 0.86} = \frac{1500}{197.8} = 7.58 \text{ AMPS}$$

TO FIND AMPERES

SINGLE PHASE:

B. When *HORSEPOWER* is known:

$$\textbf{AMPERES} = \frac{\textbf{HORSEPOWER x 746}}{\textbf{VOLTS x EFFICIENCY x POWER-FACTOR}}$$

Determine the amp-load of a single-phase, 1/2 HP, 115 volt motor. The motor has an efficiency rating of 92%, and a power-factor of 80%.

$$I = \frac{HP \times 746}{E \times \%EFF \times PF} = \frac{1/2 \times 746}{115 \times 0.92 \times 0.80} = \frac{373}{84.64}$$

I = 4.4 AMPS

C. When *KILOWATTS* are known:

$$\textbf{AMPERES} = \frac{\textbf{KILOWATTS x 1000}}{\textbf{VOLTS x POWER-FACTOR}} \quad \textbf{or} \quad I = \frac{KW \times 1000}{E \times PF}$$

A 230 Volt single-phase circuit has a 12KW power load, and operates at 84% power-factor. Determine the current.

$$I = \frac{KW \times 1000}{E \times PF} = \frac{12 \times 1000}{230 \times 0.84} = \frac{12{,}000}{193.2} = 62 \text{ AMPS}$$

D. When *KILOVOLT-AMPERE* is known:

$$\textbf{AMPERES} = \frac{\textbf{KILOVOLT-AMPERE x 1000}}{\textbf{VOLTS}} \quad \textbf{or} \quad I = \frac{KVA \times 1000}{E}$$

A 115 volt, 2 KVA, single-phase generator operating at full load will deliver 17.4 AMPERES. (Prove.)

$$I = \frac{2 \times 1000}{115} = \frac{2000}{115} = 17.4 \text{ AMPS}$$

REMEMBER:

By definition, amperes is the rate of the flow of the current.

TO FIND AMPERES

THREE PHASE:

A. When *WATTS, VOLTS, AND POWER FACTOR are known*:

$$\textbf{AMPERES} = \frac{\textbf{WATTS}}{\textbf{VOLTS} \times \textbf{POWER-FACTOR} \times 1.73}$$

$$\text{or} \qquad I = \frac{P}{E \times PF \times 1.73}$$

Determine the current when a circuit has a 1500 watt load, a power-factor of 86%, and operates from a three-phase, 230 volt source.

$$I = \frac{P}{E \times PF \times 1.73} = \frac{1500}{230 \times 0.86 \times 1.73} = \frac{1500}{342.2}$$

$$I = 4.4 \text{ AMPS}$$

B. When *HORSEPOWER* is known:

$$\textbf{AMPERES} = \frac{\textbf{HORSEPOWER} \times 746}{\textbf{VOLTS} \times \textbf{EFFICIENCY} \times \textbf{POWER-FACTOR} \times 1.73}$$

$$\text{or} \qquad I = \frac{HP \times 746}{E \times \%EFF \times PF \times 1.73}$$

Determine the amp-load of a three-phase, 1/2 HP, 230 volt motor. The motor has an efficiency rating of 92%, and a power-factor of 80%.

$$I = \frac{HP \times 746}{E \times \%EFF \times PF \times 1.73} = \frac{1/2 \times 746}{230 \times .92 \times .80 \times 1.73}$$

$$= \frac{373}{293} = 1.27 \text{ AMPS}$$

TO FIND AMPERES

THREE PHASE:

C. When *KILOWATTS are known*:

$$\textbf{AMPERES} = \frac{\textbf{KILOWATTS x 1000}}{\textbf{VOLTS x POWER-FACTOR x 1.73}}$$

$$\textbf{or} \qquad I = \frac{KW \times 1000}{E \times PF \times 1.73}$$

A 230 volt, three-phase circuit, has a 12KW power load, and operates at 84% power-factor. Determine the current.

$$I = \frac{KW \times 1000}{E \times PF \times 1.73} = \frac{12{,}000}{230 \times 0.84 \times 1.73} = \frac{12{,}000}{344.24}$$

$I = 36$ AMPS

D. When *KILOVOLT-AMPERE* is known:

$$\textbf{AMPERES} = \frac{\textbf{KILOVOLT-AMPERE x 1000}}{E \times 1.73} = \frac{KVA \times 1000}{E \times 1.73}$$

A 230 Volt, 4 KVA, three-phase generator operating at full load will deliver 10 AMPERES. (Prove.)

$$I = \frac{KVA \times 1000}{E \times 1.73} = \frac{4 \times 1000}{230 \times 1.73} = \frac{4000}{397.9}$$

$I = 10$ AMPS

NOTE: To better understand the preceding formulas:

1. TWO-PHASE CURRENT x 2 = SINGLE-PHASE CURRENT.
2. THREE-PHASE CURRENT x 1.73 = SINGLE-PHASE CURRENT
3. THE CURRENT IN THE COMMON CONDUCTOR OF A TWO-PHASE (THREE WIRE) CIRCUIT IS 141% GREATER THAN EITHER OF THE OTHER TWO CONDUCTORS OF THAT CIRCUIT.

TO FIND HORSEPOWER

DIRECT CURRENT:

$$\text{HORSEPOWER} = \frac{\text{VOLTS x AMPERES x EFFICIENCY}}{746}$$

A 12 volt motor draws a current of 8.09 amperes, and has an efficiency rating of 96%. Determine the horsepower.

$$\text{HP} = \frac{\text{E x I x \%EFF}}{746} = \frac{12 \times 8.09 \times 0.96}{746} = \frac{93.19}{746}$$

$$\text{HP} = 0.1249 = 1/8 \text{ HP}$$

SINGLE-PHASE:

$$\text{HP} = \frac{\text{VOLTS x AMPERES x EFFICIENCY x POWER FACTOR}}{746}$$

A single-phase, 115 volt (AC) motor has an efficiency rating of 92%, and a power-factor of 80%. Determine the horsepower if the amp-load is 4.4 amperes.

$$\text{HP} = \frac{\text{E x I x \%EFF x PF}}{746} = \frac{115 \times 4.4 \times 0.92 \times 0.80}{746}$$

$$\text{HP} = \frac{372.416}{746} = 0.4992 = 1/2 \text{ HP}$$

TWO-PHASE:

$$\text{HP} = \frac{\text{VOLTS x AMPERES x EFFICIENCY x POWER FACTOR x 2}}{746}$$

Determine the horsepower of a two-phase, 230 volt (AC) motor. The motor has an efficiency rating of 92%, a power-factor of 80%, and an amp-load of 1.1 amperes.

$$\text{HP} = \frac{\text{E x I x \%EFF x PF x 2}}{746} = \frac{230 \times 1.1 \times 0.92 \times 0.8 \times 2}{746}$$

$$\text{HP} = \frac{372.416}{746} = 0.4992 = 1/2 \text{ HP}$$

TO FIND HORSEPOWER

THREE-PHASE:

$$HP = \frac{\text{VOLTS} \times \text{AMPERES} \times \text{EFFICIENCY} \times \text{POWER FACTOR} \times 1.73}{746}$$

A three-phase, 460 volt motor draws a current of 52 amperes. The motor has an efficiency rating of 94%, and a power factor of 80%. Determine the horsepower.

$$HP = \frac{E \times I \times \%EFF \times PF \times 1.73}{746} = \frac{460 \times 52 \times 0.94 \times 0.80 \times 1.73}{746}$$

HP = 41.7 HP

TO FIND WATTS

The electrical power in any part of a circuit is equal to the current in that part multiplied by the voltage across that part of the circuit.

A watt is the power used when one volt causes one ampere to flow in a circuit.

One horsepower is the amount of energy required to lift 33,000 pounds, one foot, in one minute. The electrical equivalent of one horsepower is 745.6 watts. One watt is the amount of energy required to lift 44.26 pounds, one foot, in one minute. Watts is power, and power is the amount of work done in a given time.

When *VOLTS AND AMPERES* are known:

POWER (WATTS) = VOLTS x AMPERES

A 120 volt AC circuit draws a current of 5 amperes. Determine the power consumption.

$\mathbf{P = E \times I} = 120 \times 5 = 600$ WATTS

We can now determine the resistance of this circuit.

POWER = RESISTANCE x (AMPERES)²

$P = R \times I^2$ or $600 = R \times 25$ *divide both sides of equation by 25*

$\frac{600}{25} = R$ or $R = 24$ OHMS

or

$$\text{POWER} = \frac{(\text{VOLTS})^2}{\text{RESISTANCE}} \quad \text{or} \quad P = \frac{E^2}{R} \quad \text{or} \quad 600 = \frac{120^2}{R}$$

$R \times 600 = 120^2$ or $R = \frac{14{,}400}{600} = 24$ OHMS

NOTE: REFER TO THE FORMULAS OF THE OHM'S LAW CHART ON PAGE 1

TO FIND KILOWATTS

DIRECT CURRENT:

$$\textbf{KILOWATTS} = \frac{\textbf{VOLTS x AMPERES}}{\textbf{1000}}$$

A 120 volt (DC) motor draws a current of 40 amperes. Determine the kilowatts.

$$KW = \frac{E \times I}{1000} = \frac{120 \times 40}{1000} = \frac{4800}{1000} = 4.8 \text{ KW}$$

SINGLE-PHASE:

$$\textbf{KILOWATTS} = \frac{\textbf{VOLTS x AMPERES x POWER FACTOR}}{\textbf{1000}}$$

A single-phase, 115 volt (AC) motor draws a current of 20 amperes, and has a power-factor rating of 86%. Determine the kilowatts.

$$KW = \frac{E \times I \times PF}{1000} = \frac{115 \times 20 \times 0.86}{1000} = \frac{1978}{1000}$$

$$= 1.978 = 2 \text{ KW}$$

TWO-PHASE:

$$\textbf{KILOWATTS} = \frac{\textbf{VOLTS x AMPERES x POWER FACTOR x 2}}{\textbf{1000}}$$

A two-phase, 230 volt (AC) motor with a power-factor of 92%, draws a current of 55 amperes. Determine the kilowatts.

$$KW = \frac{E \times I \times PF \times 2}{1000} = \frac{230 \times 55 \times 0.92 \times 2}{1000}$$

$$KW = \frac{23.276}{1000} = 23.276 = 23 \text{ KW}$$

TO FIND KILOWATTS

THREE-PHASE:

$$\text{KILOWATTS} = \frac{\text{VOLTS} \times \text{AMPERES} \times \text{POWER FACTOR} \times 1.73}{1000}$$

A three-phase, 460 volt (AC) motor draws a current of 52 amperes, and has a power-factor rating of 80%. Determine the kilowatts.

$$\text{KW} = \frac{E \times I \times PF \times 1.73}{1000} = \frac{460 \times 52 \times 0.80 \times 1.73}{1000}$$

$$= \frac{33{,}105}{1000} = 33.105 = 33\text{KW}$$

KIRCHHOFF'S LAWS

FIRST LAW (CURRENT):

THE SUM OF THE CURRENTS ARRIVING AT ANY POINT IN A CIRCUIT MUST EQUAL THE SUM OF THE CURRENTS LEAVING THAT POINT.

SECOND LAW (VOLTAGE):

THE TOTAL VOLTAGE APPLIED TO ANY CLOSED CIRCUIT PATH IS ALWAYS EQUAL TO THE SUM OF THE VOLTAGE DROPS IN THAT PATH.

OR

THE ALGEBRAIC SUM OF ALL THE VOLTAGES ENCOUNTERED IN ANY LOOP EQUALS ZERO.

TO FIND KILOVOLT-AMPERES

SINGLE-PHASE:

$$\text{KILOVOLT-AMPERES} = \frac{\text{VOLTS} \times \text{AMPERES}}{1000}$$

A single-phase, 240 volt generator delivers 41.66 amperes at full load. Determine the kilovolt-amperes rating.

$$\text{KVA} = \frac{E \times I}{1000} = \frac{240 \times 41.66}{1000} = \frac{10{,}000}{1000} = 10\ \text{KVA}$$

TWO-PHASE:

$$\text{KILOVOLT-AMPERES} = \frac{\text{VOLTS} \times \text{AMPERES} \times 2}{1000}$$

A two-phase, 230 volt generator delivers 55 amperes. Determine the kilovolt-amperes.

$$\text{KVA} = \frac{E \times I \times 2}{1000} = \frac{230 \times 55 \times 2}{1000} = \frac{25{,}300}{1000}$$

$$= 25.3 = 25\ \text{KVA}$$

THREE-PHASE:

$$\text{KILOVOLT-AMPERES} = \frac{\text{VOLTS} \times \text{AMPERES} \times 1.73}{1000}$$

A three-phase, 460 volt generator delivers 52 amperes. Determine the kilovolt-amperes rating.

$$\text{KVA} = \frac{E \times I \times 1.73}{1000} = \frac{460 \times 52 \times 1.73}{1000} = \frac{41{,}382}{1000}$$

$$= 41.382 = 41\ \text{KVA}$$

NOTE: KVA = APPARENT POWER = POWER BEFORE USED, SUCH AS THE RATING OF A TRANSFORMER.

TO FIND CAPACITANCE

CAPACITANCE (C):

$$C = \frac{Q}{E} \quad \text{or} \quad \text{CAPACITANCE} = \frac{\text{COULOMBS}}{\text{VOLTS}}$$

Capacitance is the property of a circuit or body that permits it to store an electrical charge equal to the accumulated charge divided by the voltage. Expressed in farads.

A. To determine the total capacity of capacitors, and/or condensers connected in series.

$$\frac{1}{C_T} = \frac{1}{C_1} + \frac{1}{C_2} + \frac{1}{C_3} + \frac{1}{C_4}$$

Determine the total capacity of four each, 12 microfarad capacitors connected in series.

$$\frac{1}{C_T} = \frac{1}{C_1} + \frac{1}{C_2} + \frac{1}{C_3} + \frac{1}{C_4}$$

$$= \frac{1}{12} + \frac{1}{12} + \frac{1}{12} + \frac{1}{12} = \frac{4}{12}$$

$$\frac{1}{C_T} = \frac{4}{12} \quad \text{or} \quad C_T \times 4 = 12 \quad \text{or} \quad C_T = \frac{12}{4} = 3 \text{ microfarads}$$

B. To determine the total capacity of capacitors, and/or condensers connected in parallel.

$$C_T = C_1 + C_2 + C_3 + C_4$$

Determine the total capacity of four each, 12 microfarad capacitors connected in parallel.

$$C_T = C_1 + C_2 + C_3 + C_4$$

$$C_T = 12 + 12 + 12 + 12 = 48 \text{ microfarads}$$

A farad is the unit of capacitance of a condenser that retains one coulomb of charge with one volt difference of potential.

1 Farad = 1,000,000 Microfarads

6-DOT COLOR CODE FOR MICA AND MOLDED PAPER CAPACITORS

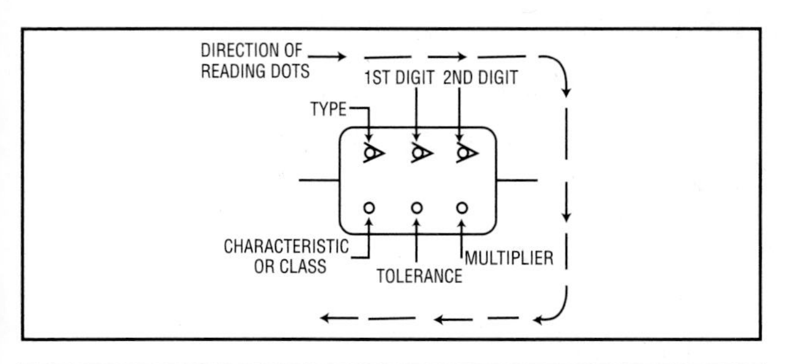

TYPE	COLOR	1ST DIGIT	2ND DIGIT	MULTIPLIER	TOLERANCE (%)	CHARACTERISTIC OR CLASS
JAN, MICA	BLACK	0	0	1	± 1	APPLIES TO TEMPERATURE COEFFICIENT OR METHODS OF TESTING
	BROWN	1	1	10	± 2	
	RED	2	2	100	± 3	
	ORANGE	3	3	1,000	± 4	
	YELLOW	4	4	10,000	± 5	
	GREEN	5	5	100,000	± 6	
	BLUE	6	6	1,000,000	± 7	
	VIOLET	7	7	10,000,000	± 8	
	GRAY	8	8	100,000,000	± 9	
ETA, MICA	WHITE	9	9	1,000,000,000		
	GOLD			.1	± 10	
MOLDED PAPER	SILVER			.01	± 20	
	BODY					

MAXIMUM PERMISSIBLE CAPACITOR KVAR FOR USE WITH OPEN-TYPE THREE-PHASE SIXTY-CYCLE INDUCTION MOTORS

	3600 RPM		1800 RPM		1200 RPM	
MOTOR RATING HP	MAXIMUM CAPACITOR RATING KVAR	REDUCTION IN LINE CURRENT %	MAXIMUM CAPACITOR RATING KVAR	REDUCTION IN LINE CURRENT %	MAXIMUM CAPACITOR RATING KVAR	REDUCTION IN LINE CURRENT %
10	3	10	3	11	3.5	14
15	4	9	4	10	5	13
20	5	9	5	10	6.5	12
25	6	9	6	10	7.5	11
30	7	8	7	9	9	11
40	9	8	9	9	11	10
50	12	8	11	9	13	10
60	14	8	14	8	15	10
75	17	8	16	8	18	10
100	22	8	21	8	25	9
125	27	8	26	8	30	9
150	32.5	8	30	8	35	9
200	40	8	37.5	8	42.5	9

	900 RPM		720 RPM		600 RPM	
MOTOR RATING HP	MAXIMUM CAPACITOR RATING KVAR	REDUCTION IN LINE CURRENT %	MAXIMUM CAPACITOR RATING KVAR	REDUCTION IN LINE CURRENT %	MAXIMUM CAPACITOR RATING KVAR	REDUCTION IN LINE CURRENT %
10	5	21	6.5	27	7.5	31
15	6.5	18	8	23	9.5	27
20	7.5	16	9	21	12	25
25	9	15	11	20	14	23
30	10	14	12	18	16	22
40	12	13	15	16	20	20
50	15	12	19	15	24	19
60	18	11	22	15	27	19
75	21	10	26	14	32.5	18
100	27	10	32.5	13	40	17
125	32.5	10	40	13	47.5	16
150	37.5	10	47.5	12	52.5	15
200	47.5	10	60	12	65	14

NOTE: If capacitors of a lower rating than the values given in the table are used, the percentage reduction in line current given in the table should be reduced proportionately.

POWER-FACTOR CORRECTION

TABLE VALUES x KW OF CAPACITORS NEEDED TO CORRECT FROM EXISTING TO DESIRED POWER FACTOR

EXISTING POWER FACTOR %	CORRECTED POWER FACTOR					
	100%	95%	90%	85%	80%	75%
50	1.732	1.403	1.247	1.112	0.982	0.850
52	1.643	1.314	1.158	1.023	0.893	0.761
54	1.558	1.229	1.073	0.938	0.808	0.676
55	1.518	1.189	1.033	0.898	0.768	0.636
56	1.479	1.150	0.994	0.859	0.729	0.597
58	1.404	1.075	0.919	0.784	0.654	0.522
60	1.333	1.004	0.848	0.713	0.583	0.451
62	1.265	0.936	0.780	0.645	0.515	0.383
64	1.201	0.872	0.716	0.581	0.451	0.319
65	1.168	0.839	0.683	0.548	0.418	0.286
66	1.139	0.810	0.654	0.519	0.389	0.257
68	1.078	0.749	0.593	0.458	0.328	0.196
70	1.020	0.691	0.535	0.400	0.270	0.138
72	0.964	0.635	0.479	0.344	0.214	0.082
74	0.909	0.580	0.424	0.289	0.159	0.027
75	0.882	0.553	0.397	0.262	0.132	
76	0.855	0.526	0.370	0.235	0.105	
78	0.802	0.473	0.317	0.182	0.052	
80	0.750	0.421	0.265	0.130		
82	0.698	0.369	0.213	0.078		
84	0.646	0.317	0.161			
85	0.620	0.291	0.135			
86	0.594	0.265	0.109			
88	0.540	0.211	0.055			
90	0.485	0.156				
92	0.426	0.097				
94	0.363	0.034				
95	0.329					

TYPICAL PROBLEM: With a load of 500 KW at 70% power factor, it is desired to find the KVA of capacitors required to correct the power factor to 85%

SOLUTION: From the table, select the multiplying factor 0.400 corresponding to the existing 70%, and the corrected 85% power factor.
0.400 x 500 = 200 KVA of capacitors required.

TO FIND INDUCTANCE

INDUCTANCE (L):

Inductance is the production of magnetization of electrification in a body by the proximity of a magnetic field or electric charge, or of the electric current in a conductor by the variation of the magnetic field in its vicinity. Expressed in Henrys.

A. To find the total inductance of coils connected in series.

$$\mathbf{L_T = L_1 + L_2 + L_3 + L_4}$$

Determine the total inductance of four coils connected in series. Each coil has an inductance of four Henrys.

$$L_T = L_1 + L_2 + L_3 + L_4$$

$$= 4 + 4 + 4 + 4 = 16 \text{ Henrys}$$

B. To find the total inductance of coils connected in parallel.

$$\mathbf{\frac{1}{L_T} = \frac{1}{L_1} + \frac{1}{L_2} + \frac{1}{L_3} + \frac{1}{L_4}}$$

Determine the total inductance of four coils connected in parallel. Each coil has an inductance of four Henrys.

$$\frac{1}{L_T} = \frac{1}{L_1} + \frac{1}{L_2} + \frac{1}{L_3} + \frac{1}{L_4}$$

$$\frac{1}{L_T} = \frac{1}{4} + \frac{1}{4} + \frac{1}{4} + \frac{1}{4}$$

$$\frac{1}{L_T} = \frac{4}{4} \text{ OR } L_T \times 4 = 1 \times 4 \text{ OR } L_T = \frac{4}{4} = 1 \text{ Henry}$$

An induction coil is a device, consisting of two concentric coils and an interrupter, that changes a low steady voltage into a high intermittent alternating voltage by electromagnetic induction. Most often used as a spark coil.

TO FIND IMPEDANCE

IMPEDANCE (Z):

Impedance is the total opposition to an alternating current presented by a circuit. Expressed in OHMS.

A. When *VOLTS AND AMPERES* are known:

$$\textbf{IMPEDANCE} = \frac{\textbf{VOLTS}}{\textbf{AMPERES}} \quad \textbf{OR} \quad Z = \frac{E}{I}$$

Determine the impedance of a 120 volt A-C circuit that draws a current of four amperes.

$$Z = \frac{E}{I} = \frac{120}{4} = 30 \text{ OHMS}$$

B. When *RESISTANCE AND REACTANCE* are known:

$$Z = \sqrt{\textbf{RESISTANCE}^2 + \textbf{REACTANCE}^2} = \sqrt{R^2 + X^2}$$

Determine the impedance of an A-C circuit when the resistance is 6 OHMS, and the reactance is 8 OHMS.

$$Z = \sqrt{R^2 + X^2} = \sqrt{36 + 64} = \sqrt{100} = 10 \text{ OHMS}$$

C. When *RESISTANCE, INDUCTIVE REACTANCE, AND CAPACITIVE REACTANCE* are known:

$$Z = \sqrt{R^2 + (X_L - X_C)^2}$$

Determine the impedance of an A-C circuit which has a resistance of 6 OHMS, an inductive reactance of 18 OHMS, and a capacitive reactance of 10 OHMS.

$$Z = \sqrt{R^2 + (X_L - X_C)^2}$$

$$= \sqrt{6^2 + (18 - 10)^2} = \sqrt{6^2 + (8)^2}$$

$$= \sqrt{36 + 64} = \sqrt{100} = 10 \text{ OHMS}$$

TO FIND REACTANCE

REACTANCE (X):

Reactance in a circuit is the opposition to an alternating current caused by inductance and capacitance, equal to the difference between capacitive and inductive reactance. Expressed in OHMS.

A. INDUCTIVE REACTANCE X_L

Inductive reactance is that element of reactance in a circuit caused by self-inductance.

$$\mathbf{X_L = 2 \times 3.1416 \times FREQUENCY \times INDUCTANCE}$$

$$= 6.28 \times F \times L$$

Determine the reactance of a four-Henry coil on a 60 cycle, A-C circuit.

$$\mathbf{X_L = 6.28 \times F \times L} = 6.28 \times 60 \times 4 = 1507 \text{ OHMS}$$

B. CAPACITIVE REACTANCE X_C

Capacitive reactance is that element of reactance in a circuit caused by capacitance.

$$\mathbf{X_C = \frac{1}{2 \times 3.1416 \times FREQUENCY \times CAPACITANCE}}$$

$$= \frac{1}{6.28 \times F \times C}$$

Determine the reactance of a four microfarad condenser on a 60 cycle, A-C circuit.

$$\mathbf{X_C = \frac{1}{6.28 \times F \times C}} = \frac{1}{6.28 \times 60 \times .000004}$$

$$= \frac{1}{0.0015072} = 663 \text{ OHMS}$$

A HENRY is a unit of inductance, equal to the inductance of a circuit in which the variation of a current at the rate of one ampere per second induces an electromotive force of one volt.

RESISTOR COLOR CODE

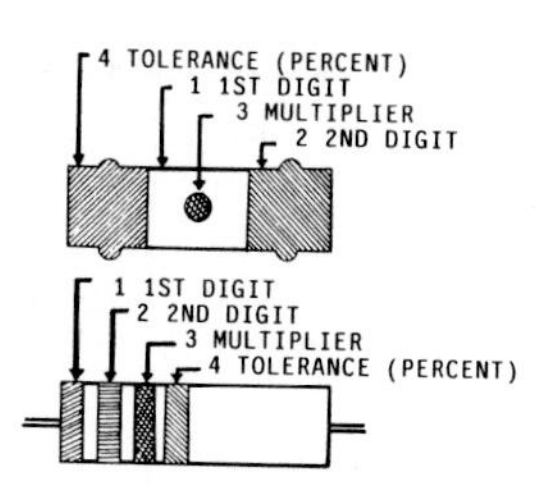

COLOR	1ST DIGIT	2ND DIGIT	MULTIPLIER	TOLERANCE (%)
BLACK	0	0	1	
BROWN	1	1	10	
RED	2	2	100	
ORANGE	3	3	1,000	
YELLOW	4	4	10,000	
GREEN	5	5	100,000	
BLUE	6	6	1,000,000	
VIOLET	7	7	10,000,000	
GRAY	8	8	100,000,000	
WHITE	9	9	1,000,000,000	
GOLD			.1	± 5%
SILVER			.01	± 10%
NO COLOR				± 20%

RUNNING OVERLOAD UNITS

KIND OF MOTOR	SUPPLY SYSTEM	NUMBER & LOCATION OF OVER-LOAD UNITS, SUCH AS TRIP COILS OR RELAYS
1-Phase ac or dc	2-wire, 1-phase ac or dc, ungrounded	1 in either conductor
1-Phase ac or dc	2-wire, 1-phase ac or dc, one conductor ungrounded	1 in ungrounded conductor
1-Phase ac or dc	3-wire, 1-phase ac or dc, grounded neutral	1 in either ungrounded conductor
1-Phase ac	any 3-phase	1 in ungrounded conductor
2-Phase ac	3-wire, 2-phase ac, ungrounded	2, one in each phase
2-Phase ac	3-wire, 2-phase ac, one conductor grounded	2 in ungrounded conductors
2-Phase ac	4-wire, 2-phase ac, grounded or ungrounded	2, one per phase in ungrounded conductors
2-Phase ac	5-wire, 2-phase ac, grounded neutral or ungrounded	2, one per phase in any ungrounded phase wire
3-Phase ac	any 3-phase	3, one in each phase*

* Exception: Where protected by other approved means.

MOTOR BRANCH - CIRCUIT PROTECTIVE DEVICES MAXIMUM RATING OR SETTING

	Percent of Full-Load Current			
Type of Motor	Nontime Delay Fuse*	Dual Element (Time-Delay) Fuse*	Instantaneous Trip Breaker	Inverse Time Breaker**
Single-phase motors	300	175	800	250
AC polyphase motors other than wound-rotor				
Squirrel Cage:				
Other than Design E	300	175	800	250
Design E	300	175	1100	250
Synchronous***	300	175	800	250
Wound rotor	150	150	800	150
Direct-current (constant voltage)	150	150	250	150

For certain exceptions to the values specified, see Sections 430-52 – 430-54.

* The values in the Nontime Delay Fuse column apply to Time-Delay Class CC fuses.

** The values given in the last column also cover the ratings of nonadjustable inverse time types of circuit breakers that may be modified as in Section 430-52.

*** Synchronous motors of the low-torque, low speed type (usually 450 rpm or lower), such as are used to drive reciprocating compressors, pumps, etc., that start unloaded, do not require a fuse rating or circuit-breaker setting in excess of 200% of full-load current.

FULL-LOAD CURRENT IN AMPERES
DIRECT CURRENT MOTORS

HP	90V	120V	180V	240V	500V	550V
1/4	4.0	3.1	2.0	1.6	–	–
1/3	5.2	4.1	2.6	2.0	–	–
1/2	6.8	5.4	3.4	2.7	–	–
3/4	9.6	7.6	4.8	3.8	–	–
1	12.2	9.5	6.1	4.7	–	–
1-1/2	–	13.2	8.3	6.6	–	–
2	–	17	10.8	8.5	–	–
3	–	25	16	12.2	–	–
5	–	40	27	20	–	–
7-1/2	–	58	–	29	13.6	12.2
10	–	76	–	38	18	16
15	–	–	–	55	27	24
20	–	–	–	72	34	31
25	–	–	–	89	43	38
30	–	–	–	106	51	46
40	–	–	–	140	67	61
50	–	–	–	173	83	75
60	–	–	–	206	99	90
75	–	–	–	255	123	111
100	–	–	–	341	164	148
125	–	–	–	425	205	185
150	–	–	–	506	246	222
200	–	–	–	675	330	294

These values of full-load currents* are for motors running at base speed.
* These are average dc quantities.

DIRECT CURRENT MOTORS

TERMINAL MARKINGS:

Terminal markings are used to tag terminals to which connections are to be made from outside circuits.

Facing the end opposite the drive (commutator end) the standard direction of shaft rotation is counter-clockwise.

A-1 and A-2 indicate armature leads.
S-1 and S-2 indicate series-field leads.
F-1 and F-2 indicate shunt-field leads.

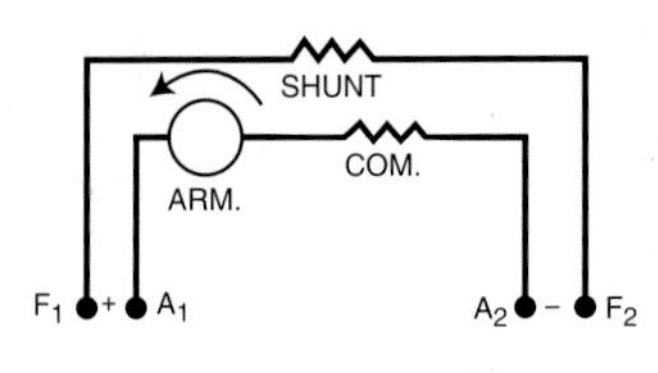

SHUNT WOUND MOTORS

To change rotation, reverse either armature leads or shunt leads. Do not reverse both armature and shunt leads.

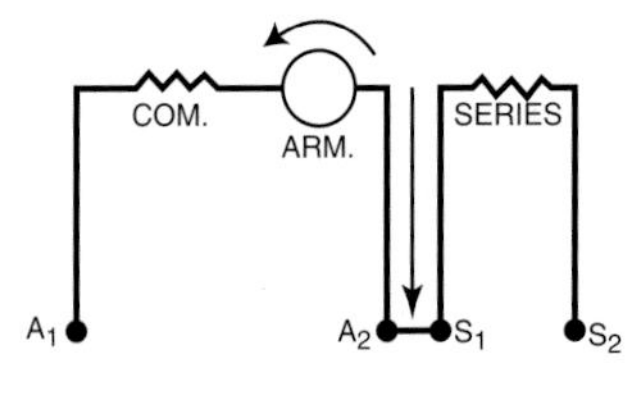

SERIES WOUND MOTORS

To change rotation, reverse either armature leads or series leads. Do not reverse both armature and series leads.

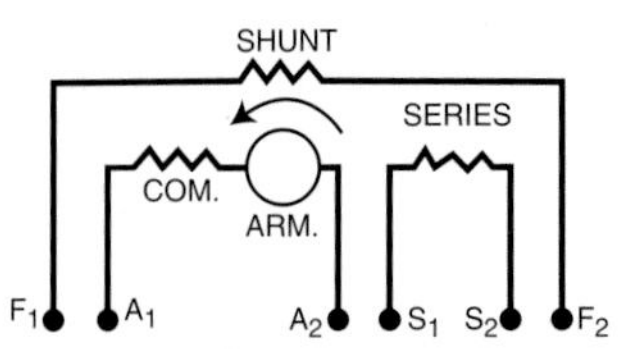

COMPOUND WOUND MOTORS

To change rotation, reverse either armature leads or both the series and shunt leads. Do not reverse all three sets of leads.

Note: Standard rotation for D.C. Generator is clockwise.

FULL-LOAD CURRENT IN AMPERES
SINGLE-PHASE ALTERNATING CURRENT MOTORS

HP	115V	200V	208V	230V
1/6	4.4	2.5	2.4	2.2
1/4	5.8	3.3	3.2	2.9
1/3	7.2	4.1	4.0	3.6
1/2	9.8	5.6	5.4	4.9
3/4	13.8	7.9	7.6	6.9
1	16	9.2	8.8	8.0
1-1/2	20	11.5	11	10
2	24	13.8	13.2	12
3	34	19.6	18.7	17
5	56	32.2	30.8	28
7-1/2	80	46	44	40
10	100	57.5	55	50

The voltages listed are rated motor voltages. The listed currents are for system voltage ranges of 110 to 120 and 220 to 240.

SINGLE-PHASE USING STANDARD THREE-PHASE STARTER

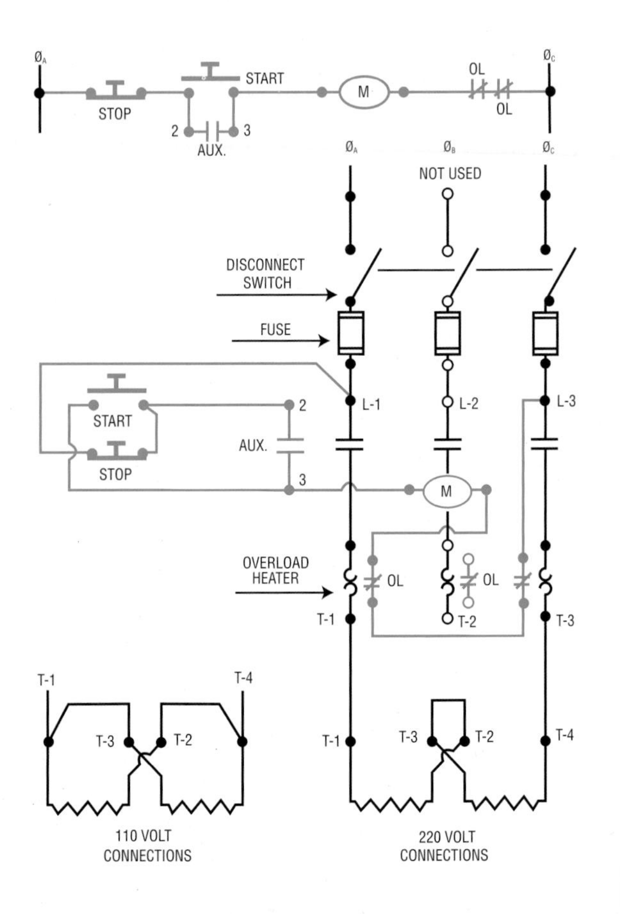

SINGLE PHASE MOTORS

SPLIT-PHASE----SQUIRREL CAGE----DUAL-VOLTAGE

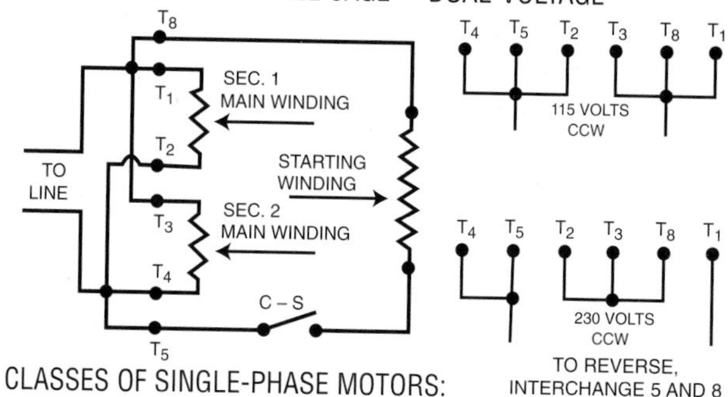

CLASSES OF SINGLE-PHASE MOTORS:

1. SPLIT-PHASE
   - A. CAPACITOR-START
   - B. REPULSION-START
   - C. RESISTANCE-START
   - D. SPLIT-CAPACITOR
2. COMMUTATOR
   - A. REPULSION
   - B. SERIES

TERMINAL COLOR MARKING:

T_1 BLUE	T_3 ORANGE	T_5 BLACK
T_2 WHITE	T_4 YELLOW	T_8 RED

NOTE: Split-phase motors are usually fractional horsepower. The majority of electric motors used in washing machines, refrigerators, etc. are of the split-phase type.

To change the speed of a split-phase motor, the number of poles must be changed.

1. Addition of running winding
2. Two starting windings, and two running windings
3. Consequent pole connections

SINGLE PHASE MOTORS

SPLIT-PHASE----SQUIRREL CAGE

A. RESISTANCE START:

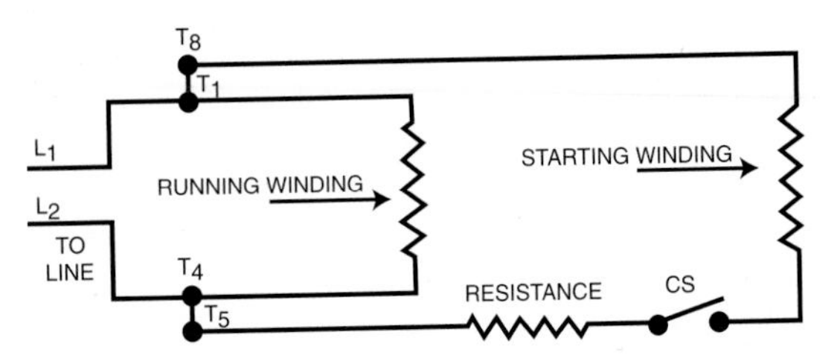

Centrifugal switch (CS) opens after reaching 75% of normal speed.

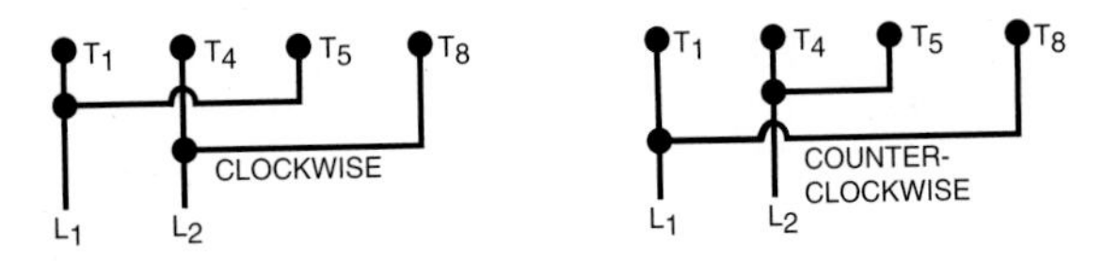

B. CAPACITOR START:

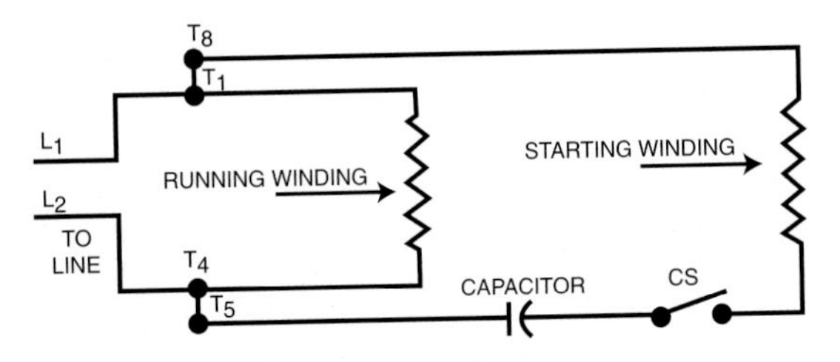

Note: 1. A resistance start motor has a resistance connected in series with the starting winding.

2. The capacitor start motor is employed where a high starting torque is required.

FULL-LOAD CURRENT
THREE-PHASE ALTERNATING CURRENT MOTORS

	Induction Type Squirrel-Cage and Wound-Rotor Amperes							Synchronous Type Unity Power Factor* Amperes			
HP	115 Volts	200 Volts	208 Volts	230 Volts	460 Volts	575 Volts	2300 Volts	230 Volts	460 Volts	575 Volts	2300 Volts
1/2	4.4	2.5	2.4	2.2	1.1	0.9	-	-	-	-	-
3/4	6.4	3.7	3.5	3.2	1.6	1.3	-	-	-	-	-
1	8.4	4.8	4.6	4.2	2.1	1.7	-	-	-	-	-
1 1/2	12.0	6.9	6.6	6.0	3.0	2.4	-	-	-	-	-
2	13.6	7.8	7.5	6.8	3.4	2.7	-	-	-	-	-
3	-	11.0	10.6	9.6	4.8	3.9	-	-	-	-	-
5	-	17.5	16.7	15.2	7.6	6.1	-	-	-	-	-
7 1/2	-	25.3	24.2	22	11	9	-	-	-	-	-
10	-	32.2	30.8	28	14	11	-	-	-	-	-
15	-	48.3	46.2	42	21	17	-	-	-	-	-
20	-	62.1	59.4	54	27	22	-	-	-	-	-
25	-	78.2	74.8	68	34	27	-	53	26	21	-
30	-	92	88	80	40	32	-	63	32	26	-
40	-	120	114	104	52	41	-	83	41	33	-
50	-	150	143	130	65	52	-	104	52	42	-
60	-	177	169	154	77	62	16	123	61	49	12
75	-	221	211	192	96	77	20	155	78	62	15
100	-	285	273	248	124	99	26	202	101	81	20
125	-	359	343	312	156	125	31	253	126	101	25
150	-	414	396	360	180	144	37	302	151	121	30
200	-	552	528	480	240	192	49	400	201	161	40
250	-	-	-	-	302	242	60	-	-	-	-
300	-	-	-	-	361	289	72	-	-	-	-
350	-	-	-	-	414	336	83	-	-	-	-
400	-	-	-	-	477	382	95	-	-	-	-
450	-	-	-	-	515	412	103	-	-	-	-
500	-	-	-	-	590	472	118	-	-	-	-

The voltages listed are rated motor voltages. The currents listed shall be permitted for system voltage ranges of 110 to 120, 220 to 240, 440 to 480, and 550-600 volts.

* For 90 and 80 percent power factor, the above figures shall be multiplied by 1.1 and 1.25 respectively.

FULL-LOAD CURRENT AND OTHER DATA THREE PHASE A.C. MOTORS

MOTOR HORSEPOWER		MOTOR AMPERE	SIZE BREAKER	SIZE STARTER	HEATER AMPERE	SIZE WIRE	SIZE CONDUIT
1/2	230V	2.2	15	00	2.530	12	3/4"
	460	1.1	15	00	1.265	12	3/4
3/4	230	3.2	15	00	3.680	12	3/4
	460	1.6	15	00	1.840	12	3/4
1	230	4.2	15	00	4.830	12	3/4
	460	2.1	15	00	2.415	12	3/4
1 1/2	230	6.0	15	00	6.900	12	3/4
	460	3.0	15	00	3.450	12	3/4
2	230	6.8	15	0	7.820	12	3/4
	460	3.4	15	00	3.910	12	3/4
3	230	9.6	20	0	11.040	12	3/4
	460	4.8	15	0	5.520	12	3/4
5	230	15.2	30	1	17.480	12	3/4
	460	7.6	15	0	8.740	12	3/4
7 1/2	230	22	45	1	25.300	10	3/4
	460	11	20	1	12.650	12	3/4
10	230	28	60	2	32.200	10	3/4
	460	14	30	1	16.100	12	3/4
15	230	42	70	2	48.300	6	1
	460	21	40	2	24.150	10	3/4
20	230	54	100	3	62.100	4	1
	460	27	50	2	31.050	10	3/4
25	230	68	100	3	78.200	4	1 1/2
	460	34	50	2	39.100	8	1
30	230	80	125	3	92.000	3	1 1/2
	460	40	70	3	46.000	8	1
40	230	104	175	4	119.600	1	1 1/2
	460	52	100	3	59.800	6	1
50	230	130	200	4	149.500	00	2
	460	65	150	3	74.750	4	1 1/2

* Overcurrent device may have to be increased due to starting current and load conditions. See NEC 430-52, Table 430-52. Wire size based on 75°C terminations and 75°C insulation.

** Overload heater must be based on motor nameplate & sized per NEC 430-32.

*** Conduit size based on Rigid Metal Conduit with some spare capacity. For minimum size & other conduit types, see NEC Appendix C, or *UGLY'S pages 75 - 95.*

FULL-LOAD CURRENT AND OTHER DATA THREE PHASE A.C. MOTORS

MOTOR HORSEPOWER		MOTOR AMPERE	SIZE BREAKER	SIZE STARTER	HEATER AMPERE	SIZE WIRE	SIZE CONDUIT
60	230V	154	250	5	177.10	000	2"
	460	77	200	4	88.55	3	1 1/2
75	230	192	300	5	220.80	250kcmil	2 1/2
	460	96	200	4	110.40	1	1 1/2
100	230	248	400	5	285.20	350kcmil	3
	460	124	200	4	142.60	1/0	2
125	230	312	500	6	358.80	600kcmil	3 1/2
	460	156	250	5	179.40	000	2
150	230	360	600	6	414.00	700kcmil	4
	460	180	300	5	207.00	0000	2 1/2

NOTE:

1. Wire & conduit size will vary depending on type of insulation & termination listing.
2. The preceding calculations apply to induction type, squirrel-cage, and wound-rotor motors only.
3. The voltages listed are rated motor voltages; corresponding nominal system voltages are 220V to 240V, and 440V to 480V.
4. Hertz: Preferred terminology for cycles per second.
5. Form coil: Coil made with rectangular or square wire.
6. Mush coil: Coil made with round wire.
7. Slip: Percentage difference between synchronous and operating speeds.
8. Synchronous speed: Maximum speed for A.C. motors or (Frequency x 120) / Poles.
9. Full load speed: Speed at which rated horsepower is developed.
10. Poles: Number of magnetic poles set up inside the motor by the placement and connection of the windings.

THREE PHASE A.C. MOTOR WINDINGS AND CONNECTIONS

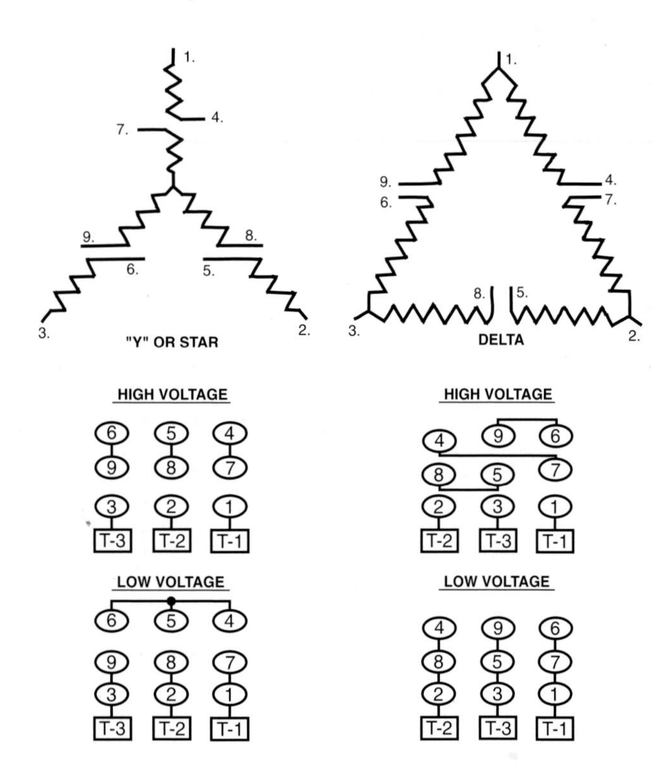

Note:
1. The most important part of any motor is the name plate. Check the data given on the plate before making the connections.
2. To change rotation direction of 3 phase motor, swap any 2 T-leads.

THREE WIRE STOP-START STATION

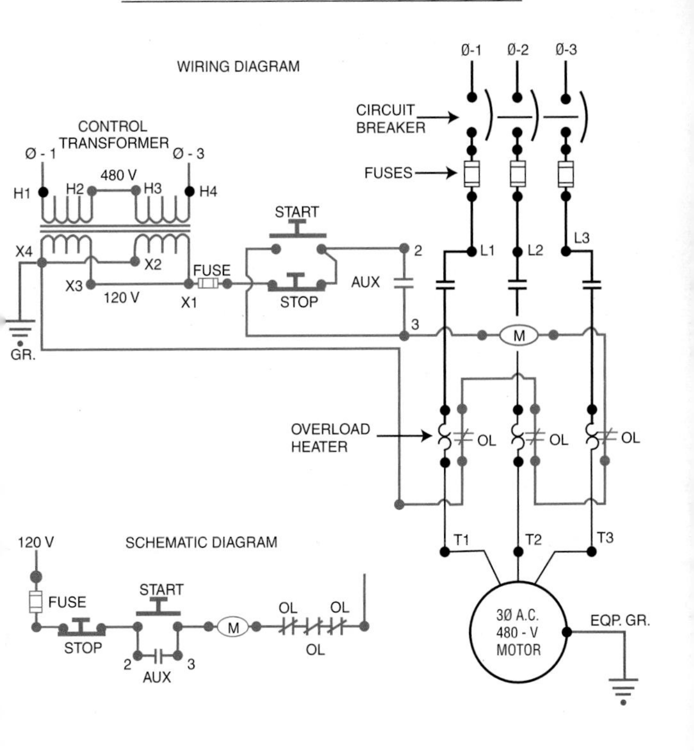

Note: Controls and motor are of different voltages..

TWO THREE WIRE STOP-START STATIONS

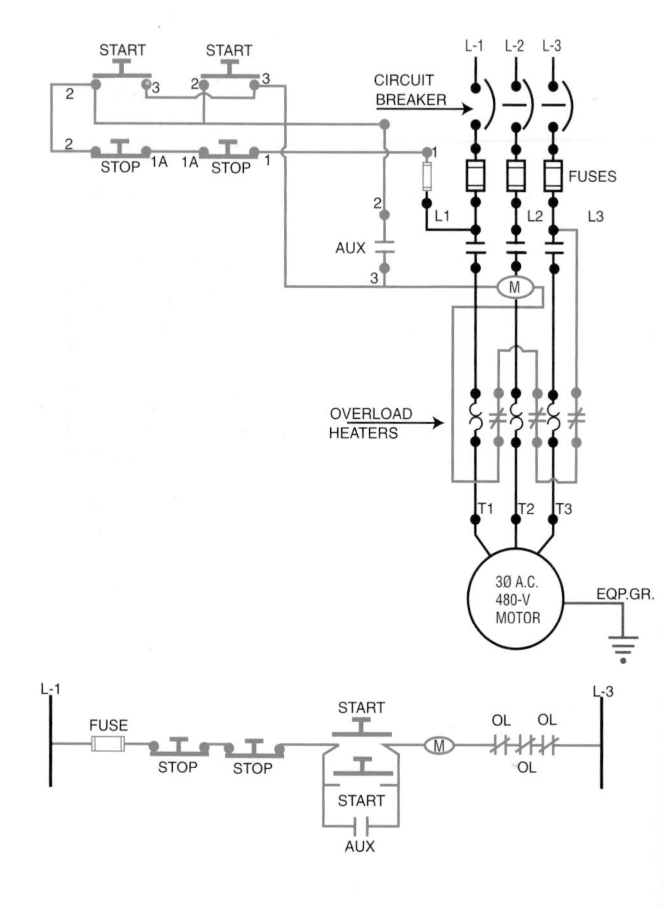

Note: Controls and motor are of the same voltage.
If Low Voltage controls are used, see UGLY'S page 43 for control transformer connections.

HAND OFF AUTOMATIC CONTROL

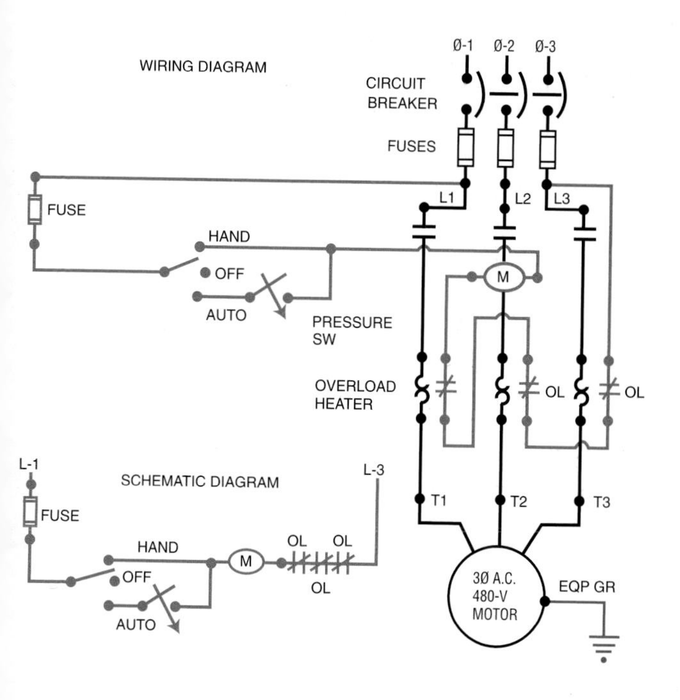

Note: Controls and motor are of the same voltage.
If Low Voltage controls are used, see UGLY'S page 43 for control transformer connections.

JOGGING WITH CONTROL RELAY

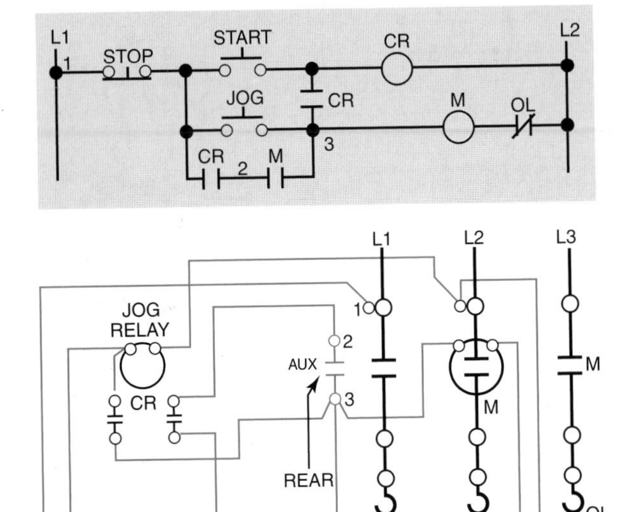

Jogging circuits are used when machines must be operated momentarily for inching (as in set-up or maintenance). The jog circuit allows the starter to be energized only as long as the jog button is depressed.

TRANSFORMER CALCULATIONS

To better understand the following formulas, review the rule of transposition in equations.

A multiplier may be removed from one side of an equation by making it a divisor on the other side; or a divisor may be removed from one side of an equation by making it a multiplier on the other side.

1. VOLTAGE AND CURRENT: PRIMARY (p) AND SECONDARY (s)

POWER (p) = POWER (s) or $Ep \times Ip = Es \times Is$

A. $Ep = \frac{Es \times Is}{Ip}$

B. $Ip = \frac{Es \times Is}{Ep}$

C. $\frac{Ep \times Ip}{Es} = Is$

D. $\frac{Ep \times Ip}{Is} = Es$

2. VOLTAGE AND TURNS IN COIL:

VOLTAGE (p) x TURNS (s) = VOLTAGE (s) x TURNS (p)

or

$$Ep \times Ts = Es \times Tp$$

A. $Ep = \frac{Es \times Tp}{Ts}$

B. $Ts = \frac{Es \times Tp}{Ep}$

C. $\frac{Ep \times Ts}{Es} = Tp$

D. $\frac{Ep \times Ts}{Tp} = Es$

3. AMPERES AND TURNS IN COIL:

AMPERES (p) x TURNS (p) = AMPERES (s) x TURNS (s)

or

$$Ip \times Tp = Is \times Ts$$

A. $Ip = \frac{Is \times Ts}{Tp}$

B. $Tp = \frac{Is \times Ts}{Ip}$

C. $\frac{Ip \times Tp}{Is} = Ts$

D. $\frac{Ip \times Tp}{Ts} = Is$

VOLTAGE DROP CALCULATIONS
INDUCTANCE NEGLIGIBLE

Vd = Voltage Drop
I = Current in Conductor (Amperes)
L = One-way Length of Circuit (Ft.)
Cm = Cross Section Area of Conductor (Circular Mils)
K = Resistance in ohms of one circular mil foot of conductor

K = 12.9 for Copper Conductors @75°C
K = 21.2 for Aluminum Conductors @75°C

NOTE: K value changes with temperature.
See NEC chapter 9, Table 8, Notes

* TWO-WIRE SINGLE PHASE CIRCUITS:

$$Vd = \frac{2K \times L \times I}{Cm} \quad \text{or} \quad Cm = \frac{2K \times L \times I}{Vd}$$

* THREE-WIRE SINGLE PHASE CIRCUITS:

$$Vd = \frac{2K \times L \times I}{Cm} \quad \text{or} \quad Cm = \frac{2K \times L \times I}{Vd}$$

* THREE-WIRE THREE PHASE CIRCUITS:

$$Vd = \frac{1.73K \times L \times I}{Cm} \quad \text{or} \quad Cm = \frac{1.73K \times L \times I}{Vd}$$

Note: Always check ampacity tables for conductors selected

Refer to UGLY'S pages 65 - 70 for conductor size, type & ampacity.

ALTERNATE VOLTAGE DROP CALCULATIONS

(Courtesy of Cooper Bussmann)

How to figure volt loss (drop):

Multiply the **Distance** (length in feet of one wire) by the **Current** (expressed in amperes) by the **Multiplier** shown in Table A (see next page) for the kind of current and the size of wire to be used, by one over the number of conductors per phase. Then place a decimal point in front of the last 6 digits.

This results in the volt loss to be expected on that circuit.

Example: No. 6 copper wire in 180 feet of iron conduit - 3 phase, 40 amp. load at 80% power factor.

Multiply feet by amperes: 180 x 40 = 7200

Multiply this number by the multiplier from Table A for No. 6 wire, three phase at 80% power factor. 7200 x *745* = 5364000.

Multiply by $\frac{1}{\#/\text{phase}}$ $5364000 \times \frac{1}{1} = 5364000$

Place the decimal point 6 places to the left. 5.364 volts expected volt loss.

(For a 240 volt circuit, the % voltage drop is $\frac{5.364}{240} \times 100$ or 2.23%.)

This table takes into consideration reactance on AC circuits as well as resistance of the wire. Re member on short runs to verify that the size and type of wire indicated has sufficient ampere capacity.

How to select size of wire:

Multiply the **Distance** (length in feet of one wire) by the **Current** (expressed in amperes), by one over the number of conductors per phase.

Divide that figure into the permissible **Volt Loss** multiplied by 1,000,000.

In Table A, look in the column applying to the type of current and power factor for the figure nearest to, but not above your result. Then follow that row to the left and this is the wire size needed.

Example:

Copper in 180 feet of iron conduit - 3 phase, 40 amp. load at 80% power factor - volt loss from local code equals 5.5 volts.

Multiply feet by amperes by $\frac{1}{\#/\text{phase}}$: $180 \times 40 \times \frac{1}{1} = 7200$

Divide (permissible volt loss multiplied by 1,000,000) by this number. $\frac{5.5 \times 1{,}000{,}000}{7200} = 764$

From Table A, select the number from the 3 phase, 80% power factor column that is nearest, but not greater than 764. This number is 745, indicating the size or wire needed is No.6.

TABLE A

MULTIPLIER FOR VOLTAGE DROP CALCULATIONS - STEEL CONDUIT

WIRE SIZE	AMPACITY TYPE T,TW (60°C WIRE)	AMPACITY TYPE RH, THWN, RHW, THW (75°C WIRE)	AMPACITY TYPE RHH, THHN, XHHW 90°C WIRE)	DIRECT CURRENT	THREE PHASE (60 CYCLE, LAGGING POWER FACTOR) 100%	THREE PHASE 90%	THREE PHASE 80%	THREE PHASE 70%	THREE PHASE 60%	SINGLE PHASE (60 CYCLE, LAGGING POWER FACTOR) 100%	SINGLE PHASE 90%	SINGLE PHASE 80%	SINGLE PHASE 70%	SINGLE PHASE 60%
14	20	20	25	6140	5369	4887	4371	3848	3322	6200	5643	5047	4444	3836
12	25	25	30	3860	3464	3169	2841	2508	2172	4000	3659	3281	2897	2508
10	30	35	40	2420	2078	1918	1728	1532	1334	2400	2214	1995	1769	1540
8	40	50	55	1528	1350	1264	1148	1026	900	1560	1460	1326	1184	1040
6	55	65	75	982	848	812	745	673	597	980	937	860	777	690
4	70	85	95	616	536	528	491	450	405	620	610	568	519	468
3	85	100	110	490	433	434	407	376	341	500	501	470	434	394
2	95	115	130	388	346	354	336	312	286	400	409	388	361	331
1	110	130	150	308	277	292	280	264	245	320	337	324	305	283
0	125	150	170	244	207	228	223	213	200	240	263	258	246	232
00	145	175	195	193	173	196	194	188	178	200	227	224	217	206
000	165	200	225	153	136	162	163	160	154	158	187	188	184	178
0000	195	230	260	122	109	136	140	139	136	126	157	162	161	157
250	215	255	290	103	93	123	128	129	128	108	142	148	149	148
300	240	285	320	86	77	108	115	117	117	90	125	133	135	135
350	260	310	350	73	67	98	106	109	109	78	113	122	126	126
400	280	335	380	64	60	91	99	103	104	70	105	114	118	120
500	320	380	430	52	50	81	90	94	96	58	94	104	109	111
600	355	420	475	43	43	75	84	898	92	50	86	97	103	106
750	400	475	535	34	36	68	78	84	88	42	79	91	97	102
1000	455	545	615	26	31	62	72	78	82	36	72	84	90	95

SHORT CIRCUIT CALCULATION

(Courtesy of Cooper Bussmann)

Basic Short-Circuit Calculation Procedure:

1. Determine transformer full-load amperes from either:
   a) Name plate
   b) Formula:

   3ø transf. $$I_{l.l.} = \frac{KVA \times 1000}{E_{L-L} \times 1.732}$$

   1ø transf. $$I_{l.l.} = \frac{KVA \times 1000}{E_{L-L}}$$

2. Find transformer multiplier.

   $$\text{Multilpier} = \frac{100}{{}^{*}\%Z_{trans}}$$

3. Determine transformer let-thru short-circuit current.**

   $$I_{S.C.} = I_{l.l.} \times \text{Multiplier}$$

4. Calculate "f" factor.

   3Ø faults $$f = \frac{1.732 \times L \times I_{3Ø}}{C \times E_{L-L}}$$

   1Ø line-to-line (L-L) faults on 1Ø Center Tapped Transformer $$f = \frac{2 \times L \times I_{L-L}}{C \times E_{L-L}}$$

   1Ø line-to-neutral (L-N) faults on 1Ø Center Tapped Transformer $$f = \frac{2 \times L \times I_{L-N}^{***}}{C \times E_{L-N}}$$

   L = length (feet) of conductor to the fault
   C = constant from Table C (page 53) for conductors & busway. For parallel runs, multiply C values by the number of conductors per phase.
   I = available short-circuit current in amperes at beginning of circuit.

5. Calculate "M" (multiplier) $$M = \frac{1}{1 + f}$$

6. Calculate the available short-circuit symmetrical RMS current at the point of fault.

   $$I_{S.C.\ sym\ RMS} = I_{S.C.} \times M$$

(Continued - next page)

(Courtesy of Cooper Bussmann)

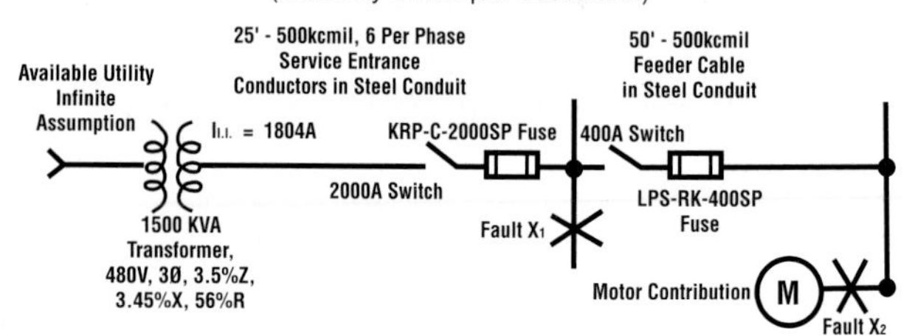

Example - Short-Circuit Calculation:

(FAULT #1)

1. $$I_{I.I.} = \frac{KVA \times 1000}{E_{L-L} \times 1.732} = \frac{1500 \times 1000}{480 \times 1.732} = 1804\ A$$

2. $$\text{Multilpier} = \frac{100}{{}^{*}\ \%Z_{trans}} = \frac{100}{3.5} = 28.57$$

3. $$I_{S.C.} = 1804 \times 28.57 = 51{,}540\ A$$

4. $$f = \frac{1.732 \times L \times I_{3Ø}}{C \times E_{L-L}} = \frac{1.73 \times 25 \times 51{,}540}{6 \times 22{,}185 \times 480} = 0.0349$$

5. $$M = \frac{1}{1 + f} = \frac{1}{1 + .0349} = .9663$$

6. $I_{S.C.\ sym\ RMS} = I_{S.C.} \times M = 51{,}540 \times .9663 = 49{,}803\ A$
   $I_{S.C.\ motor\ contrib} = 4 \times 1{,}804 = 7{,}216\ A$
   $I_{total\ S.C.\ sym\ RMS} = 49{,}803 + 7{,}216 = 57{,}019\ A$

(FAULT #2)

4. Use $I_{S.C.\ sym\ RMS}$ @ Fault X_1 to calculate "f"

$$f = \frac{1.73 \times 50 \times 49{,}803}{22{,}185 \times 480} = 0.4050$$

5. $$M = \frac{1}{1 + .4050} = .7117$$

6. $I_{S.C.\ sym\ RMS} = 49{,}803 \times .7117 = 35{,}445\ A$
   $I_{sym\ motor\ contrib} = 4 \times 1{,}804 = 7{,}216\ A$
   $I_{total\ S.C.\ sym\ RMS} = 35{,}445 + 7{,}216 = 42{,}661\ A$

(Continued - next page)

TABLE "C"

"C" VALUES for CONDUCTORS (SHORT-CIRCUIT CALCULATION)

(Courtesy of Cooper Bussmann)

AWG or MCM	Copper Three Single Conductors						Copper Three Conductor Cable					
	Steel Conduit			Nonmagnetic Conduit			Steel Conduit			Nonmagnetic Conduit		
	600V	5KV	15KV	600V	5KV	15KV	600V	5KV	15KV	600V	5KV	15KV
12	617	617	617	617	617	617	617	617	617	617	617	617
10	981	981	981	981	981	981	981	981	981	981	981	981
8	1557	1551	1557	1558	1555	1558	1559	1557	1559	1559	1558	1559
6	2425	2406	2389	2430	2417	2406	2431	2424	2414	2433	2428	2420
4	3806	3750	3695	3825	3789	3752	3830	3811	3778	3837	3823	3798
3	4760	4760	4760	4802	4802	4802	4760	4790	4760	4802	4802	4802
2	5906	5736	5574	6044	5926	5809	5989	5929	5827	6087	6022	5957
1	7292	7029	6758	7493	7306	7108	7454	7364	7188	7579	7507	7364
1/0	8924	8543	7973	9317	9033	8590	9209	9086	8707	9472	9372	9052
2/0	10755	10061	9389	11423	10877	10318	11244	11045	10500	11703	11528	11052
3/0	12843	11804	11021	13923	13048	12360	13656	13333	12613	14410	14118	13461
4/0	15082	13605	12542	16673	15351	14347	16391	15890	14813	17482	17019	16012
250	16483	14924	13643	18593	17120	15865	18310	17850	16465	19779	19352	18001
300	18176	16292	14768	20867	18975	17408	20617	20051	18318	22524	21938	20163
350	19703	17385	15678	22736	20526	18672	22646	21914	19821	24904	24126	21982
400	20565	18235	16365	24296	21786	19731	24253	23371	21042	26915	26044	23517
500	22185	19172	17492	26706	23277	21329	26980	25449	23125	30028	28712	25916
600	22965	20567	17962	28033	25203	22097	28752	27974	24896	32236	31258	27766
750	24136	21386	18888	28303	25430	22690	31050	30024	26932	32404	31338	28303
1000	25278	22539	19923	31490	28083	24887	33864	32688	29320	37197	35748	31959

(Courtesy of Cooper Bussmann)

NOTES:

* Transformer impedance (Z) helps to determine what the short circuit current will be at the transformer secondary. Transformer impedance is determined as follows: The transformer secondary is short circuited. Voltage is applied to the primary which causes full load current to flow in the secondary. This applied voltage divided by the rated primary voltage is the impedance of the transformer.

Example:

For a 480 volt rated primary, if 9.6 volts causes secondary full load current to flow through the shorted secondary, the transformer impedance is $9.6 \div 480 = .02 = 2\%Z$. In addition, U.L. listed transformers 25KVA and larger have a ± 10% impedance tolerance. Short circuit amperes can be affected by this tolerance.

** Motor Short-circuit contribution, if significant, may be added to the transformer secondary short-circuit current value as determined in Step 3. Proceed with this adjusted figure through Steps 4,5 and 6. A practical estimate of motor short-circuit contribution is to multiply the total motor current in amperes by 4.

*** The L-N fault current is higher than the L-L fault current at the secondary terminals of a single-phase center-tapped transformer. The short-circuit current available (I) for this case in Step 4 should be adjusted at the transformer terminals as follows:

At L-N center tapped transformer terminals,

$I_{L-N} = 1.5 \times I_{L-L}$ at Transformer Terminals.

SINGLE-PHASE TRANSFORMER CONNECTIONS

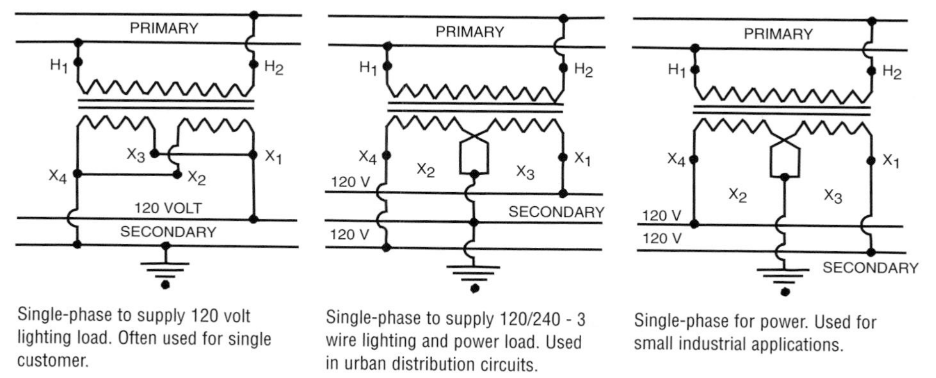

Single-phase to supply 120 volt lighting load. Often used for single customer.

Single-phase to supply 120/240 - 3 wire lighting and power load. Used in urban distribution circuits.

Single-phase for power. Used for small industrial applications.

SINGLE Ø TRANSFORMER CIRCUIT

A transformer is a stationary induction device for transferring electrical energy from one circuit to another without change of frequency. A transformer consists of two coils or windings wound upon a magnetic core of soft iron laminations, and insulated from one another.

BUCK AND BOOST TRANSFORMER CONNECTIONS

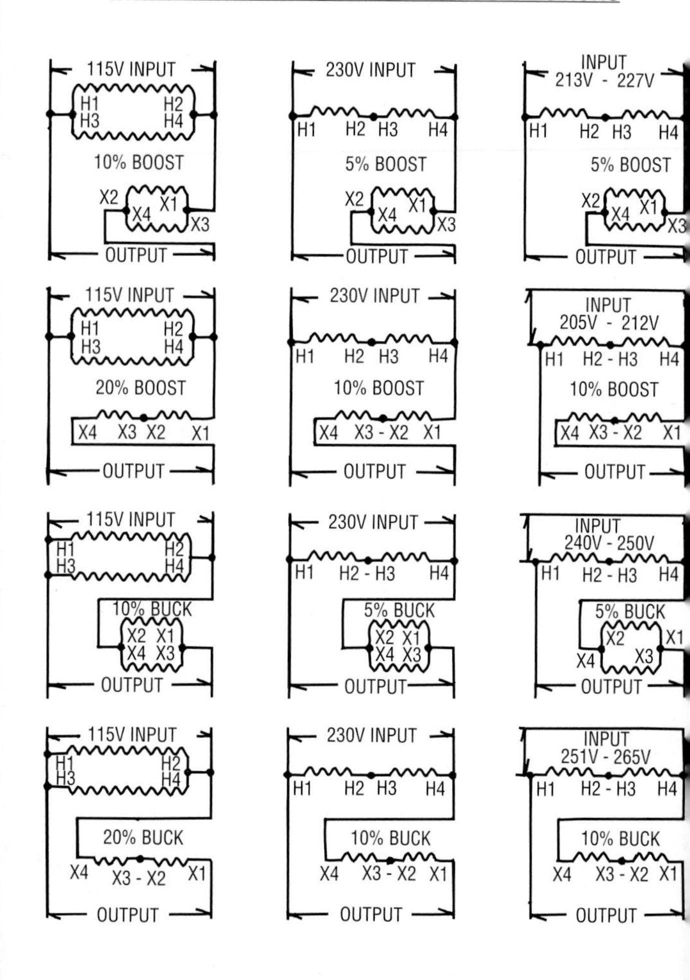

FULL LOAD CURRENTS

THREE-PHASE TRANSFORMERS VOLTAGE (LINE TO LINE)						SINGLE-PHASE TRANSFORMERS VOLTAGE					
KVA RATING	208	240	480	2400	4160	KVA RATING	120	208	240	480	2400
3	8.3	7.2	3.6	.72	.416	1	8.33	4.81	4.17	2.08	.42
6	16.7	14.4	7.2	1.44	.83	3	25.0	14.4	12.5	6.25	1.25
9	25.0	21.7	10.8	2.17	1.25	5	41.7	24.0	20.8	10.4	2.08
15	41.6	36.1	18.0	3.6	2.08	7.5	62.5	36.1	31.3	15.6	3.13
30	83.3	72.2	36.1	7.2	4.16	10	83.3	48.1	41.7	20.8	4.17
45	124.9	108.3	54.1	10.8	6.25	15	125.0	72.1	62.5	31.3	6.25
75	208.2	180.4	90.2	18.0	10.4	25	208.3	120.2	104.2	52.1	10.4
100	277.6	240.6	120.3	24.1	13.9	37.5	312.5	180.3	156.3	78.1	15.6
150	416.4	360.9	180.4	36.1	20.8	50	416.7	240.4	208.3	104.2	20.8
225	624.6	541.3	270.6	54.1	31.2	75	625.0	360.6	312.5	156.3	31.3
300	832.7	721.7	360.9	72.2	41.6	100	833.3	480.8	416.7	208.3	41.7
500	1387.9	1202.8	601.4	120.3	69.4	125	1041.7	601.0	520.8	260.4	52.1
750	2081.9	1804.3	902.1	180.4	104.1	167.5	1395.8	805.3	697.9	349.0	69.8
1000	2775.8	2405.7	1202.8	240.6	138.8	200	1666.7	961.5	833.3	416.7	83.3
1500	4163.7	3608.5	1804.3	360.9	208.2	250	2083.3	1201.9	1041.7	520.8	104.2
2000	5551.6	4811.4	2405.7	481.1	277.6	333	2775.0	1601.0	1387.5	693.8	138.8
2500	6939.5	6014.2	3007.1	601.4	347.0	500	4166.7	2403.8	2083.3	1041.7	208.3
5000	13879.0	12028.5	6014.2	1202.8	694.0						
7500	20818.5	18042.7	9021.4	1804.3	1040.9						
10000	27758.0	24057.0	12028.5	2405.7	1387.9						

Three-phase: $I = \frac{KVA \times 1000}{E \times 1.73}$ or $KVA = \frac{E \times I \times 1.73}{1000}$

Single-phase: $I = \frac{KVA \times 1000}{E}$ or $KVA = \frac{E \times I}{1000}$

THREE PHASE CONNECTIONS

WYE (STAR)

Voltage from "A", "B", or "C" to Neutral = E_{PHASE} (E_P)
Voltage between A-B, B-C, or C-A = E_{LINE} (E_L)
$I_L = I_P$, if balanced.
If unbalanced,
$I_L = \sqrt{I_A^2 + I_B^2 + I_C^2 - (I_A \times I_B) - (I_B \times I_C) - (I_C \times I_A)}$
$E_L = E_P \times 1.73$
$E_P = E_L \div 1.73$
Power =
$I_L \times E_L \times 1.73 \times$ Power Factor (cosine)
Voltamperes = $I_L \times E_L \times 1.73$

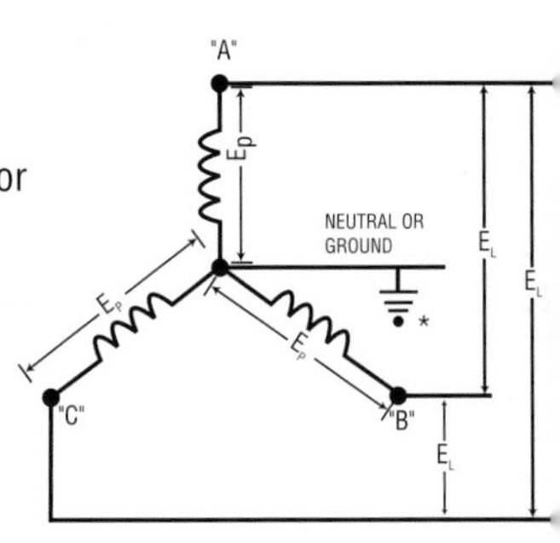

DELTA

E_{LINE} (E_L) = E_{PHASE} (E_P)
$I_{LINE} = I_P \times 1.73$
$I_{PHASE} = I_L \div 1.73$
Power =
$I_L \times E_L \times 1.73 \times$ Power Factor (cosine)
Voltamperes = $I_L \times E_L \times 1.73$

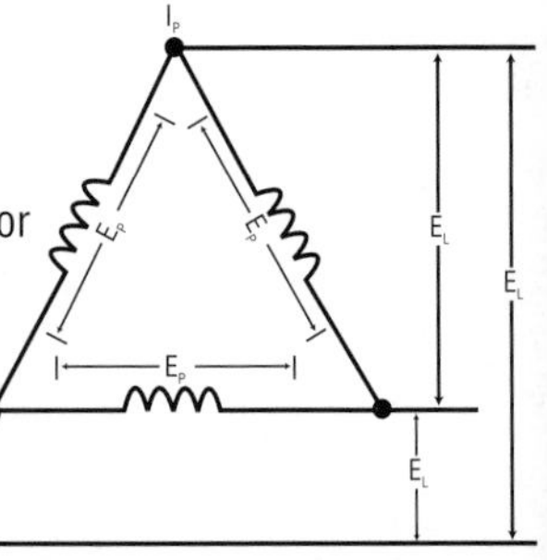

* Neutral could be ungrounded
Also see NEC A250 System Grounding Requirements

THREE-PHASE STANDARD PHASE ROTATION

TRANSFORMERS

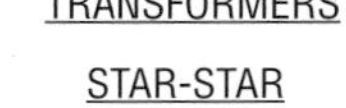

STAR-DELTA

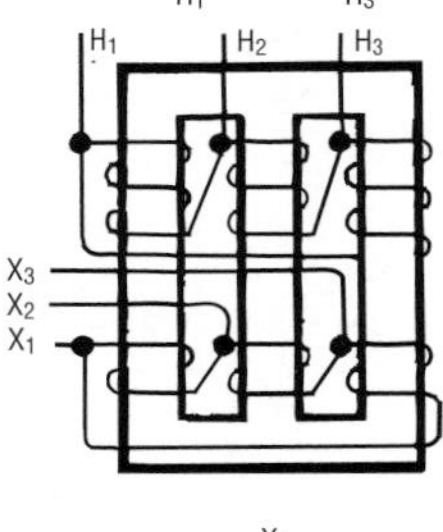

ADDITIVE POLARITY
30° ANGULAR-DISPLACEMENT

STAR-STAR

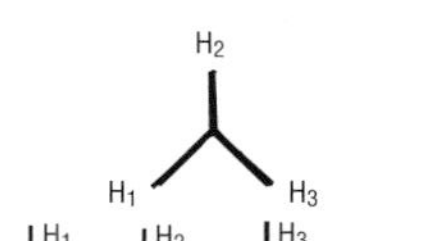

SUBTRACTIVE POLARITY
0° PHASE-DISPLACEMENT

DELTA-DELTA

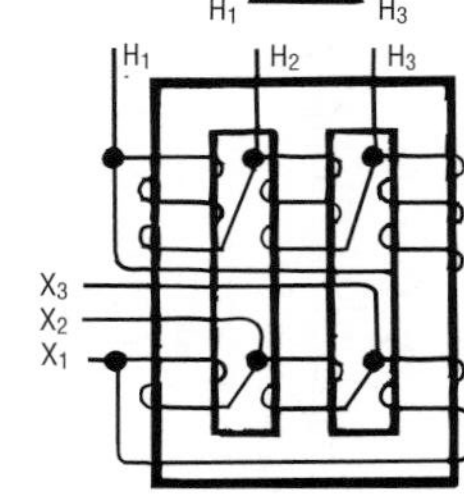
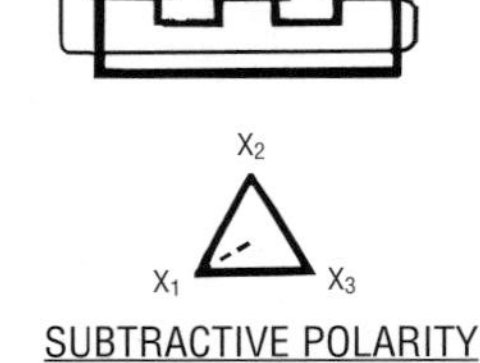

SUBTRACTIVE POLARITY
0° PHASE-DISPLACEMENT

TRANSFORMER CONNECTIONS

SERIES CONNECTIONS OF LOW VOLTAGE WINDINGS

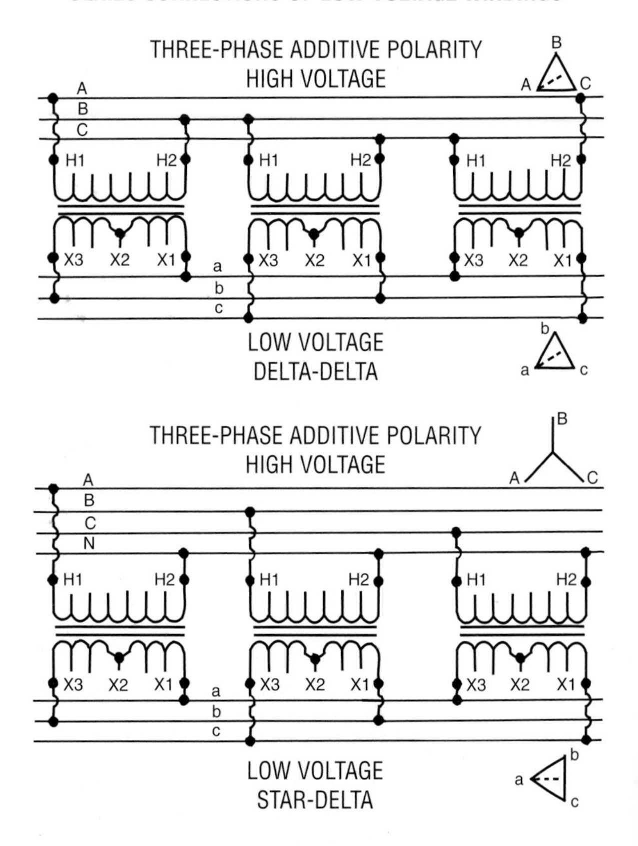

NOTE: Single-phase transformers should be thoroughly checked for impedance, polarity, and voltage ratio before installation.

TRANSFORMER CONNECTIONS

SERIES CONNECTIONS OF LOW VOLTAGE WINDINGS

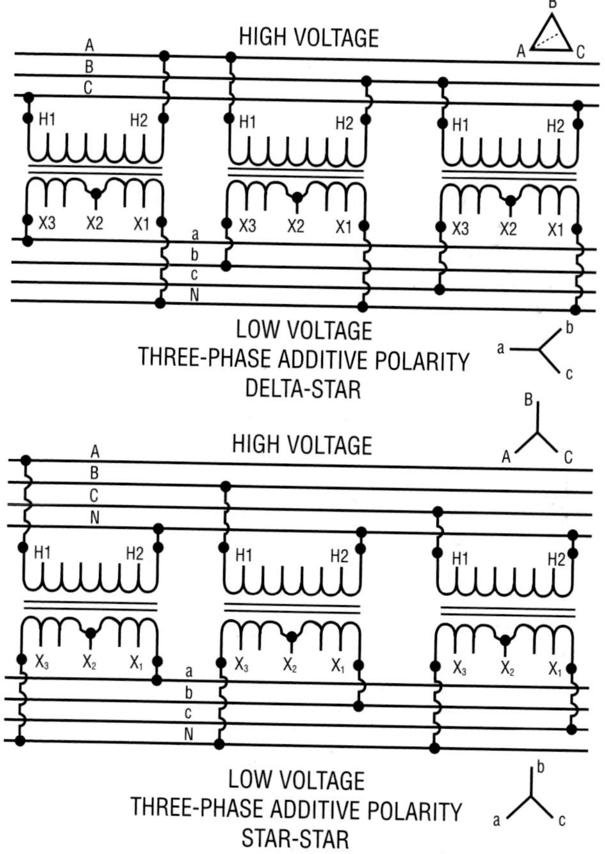

NOTE: For additive polarity the H-1 and the X-1 bushings are diagonally opposite each other.

TRANSFORMER CONNECTIONS

SERIES CONNECTIONS OF LOW VOLTAGE WINDINGS

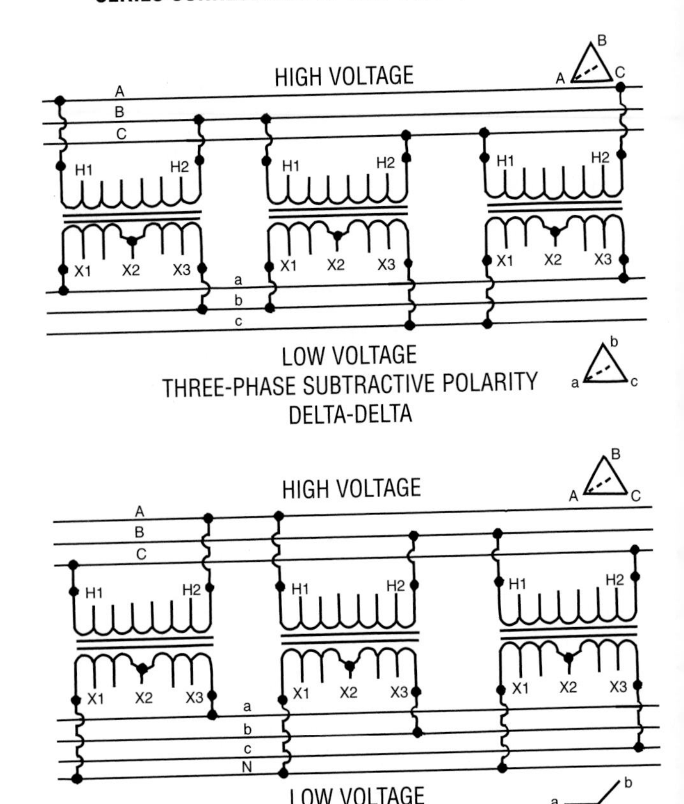

NOTE: For subtractive polarity the H-1 and the X-1 bushings are directly opposite each other.

MISCELLANEOUS WIRING DIAGRAMS

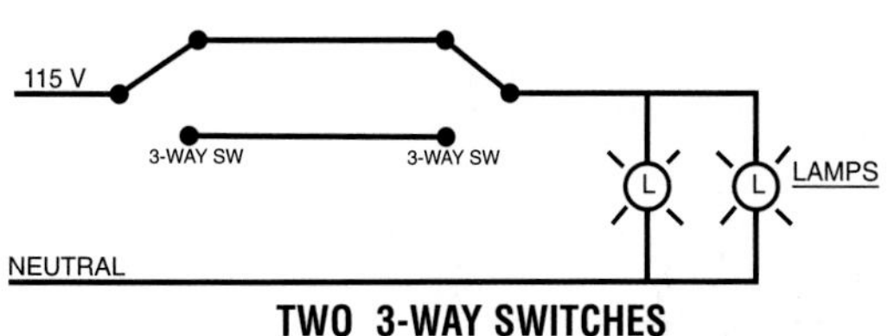

TWO 3-WAY SWITCHES

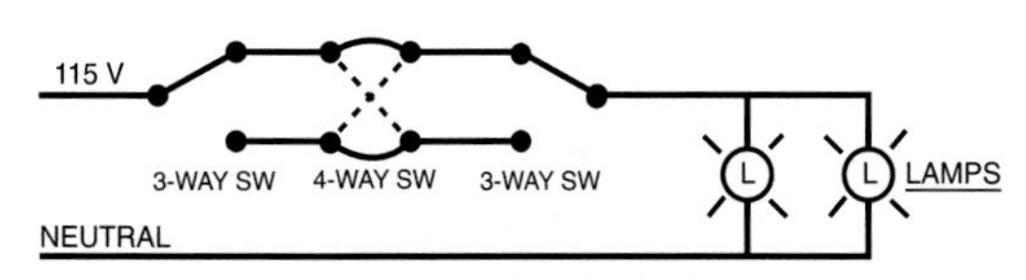

TWO 3-WAY SWITCHES
ONE 4-WAY SWITCH

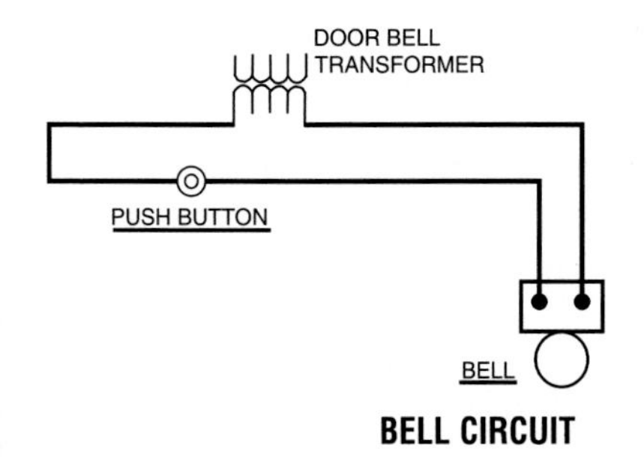

BELL CIRCUIT

MISCELLANEOUS WIRING DIAGRAMS

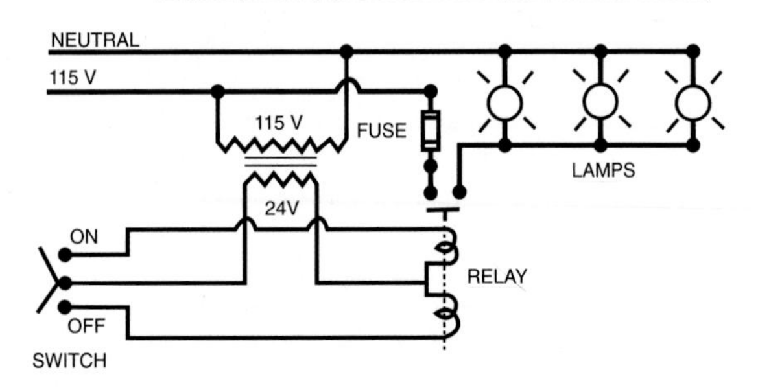

REMOTE CONTROL CIRCUIT - ONE RELAY AND ONE SWITCH

SUPPORTS FOR RIGID METAL CONDUIT

CONDUIT SIZE	DISTANCE BETWEEN SUPPORTS
1/2" - 3/4"	10 FEET
1"	12 FEET
1-1/4" - 1-1/2"	14 FEET
2" - 2-1/2"	16 FEET
3" AND LARGER	20 FEET

SUPPORT OF RIGID NONMETALLIC CONDUIT

CONDUIT SIZE	DISTANCE BETWEEN SUPPORTS
1/2" - 1"	3 FEET
1-1/4" - 2"	5 FEET
2-1/2" - 3"	6 FEET
3-1/2" - 5"	7 FEET
6"	8 FEET

For SI units: (Supports) one foot = 0.3048 meter

CONDUCTOR PROPERTIES

		Conductors				DC Resistance at 75°C (167°F)		
Size	Area	Stranding		Overall		Copper		Aluminum
AWG/ kcmil	Cir. Mils	Quan-tity	Diam. In.	Diam. In.	Area In^2	Uncoated ohm/kFT	Coated ohm/kFT	ohm/ kFT
18	1620	1	----	0.040	0.001	7.77	8.08	12.8
18	1620	7	0.015	0.046	0.002	7.95	8.45	13.1
16	2580	1	----	0.051	0.002	4.89	5.08	8.05
16	2580	7	0.019	0.058	0.003	4.99	5.29	8.21
14	4110	1	----	0.064	0.003	3.07	3.19	5.06
14	4110	7	0.024	0.073	0.004	3.14	3.26	5.17
12	6530	1	----	0.081	0.005	1.93	2.01	3.18
12	6530	7	0.030	0.092	0.006	1.98	2.05	3.25
10	10380	1	----	0.102	0.008	1.21	1.26	2.00
10	10380	7	0.038	0.116	0.011	1.24	1.29	2.04
8	16510	1	----	0.128	0.013	0.764	0.786	1.26
8	16510	7	0.049	0.146	0.017	0.778	0.809	1.28
6	26240	7	0.061	0.184	0.027	0.491	0.510	0.808
4	41740	7	0.077	0.232	0.042	0.308	0.321	0.508
3	52620	7	0.087	0.260	0.053	0.245	0.254	0.403
2	66360	7	0.097	0.292	0.067	0.194	0.201	0.319
1	83690	19	0.066	0.332	0.087	0.154	0.160	0.253
1/0	105600	19	0.074	0.372	0.109	0.122	0.127	0.201
2/0	133100	19	0.084	0.418	0.137	0.0967	0.101	0.159
3/0	167800	19	0.094	0.470	0.173	0.0766	0.0797	0.126
4/0	211600	19	0.106	0.528	0.219	0.0608	0.0626	0.100
250	----	37	0.082	0.575	0.260	0.0515	0.0535	0.0847
300	----	37	0.090	0.630	0.312	0.0429	0.0446	0.0707
350	----	37	0.097	0.681	0.364	0.0367	0.0382	0.0605
400	----	37	0.104	0.728	0.416	0.0321	0.0331	0.0529
500	----	37	0.116	0.813	0.519	0.0258	0.0265	0.0424
600	----	61	0.099	0.893	0.626	0.0214	0.0223	0.0353
700	----	61	0.107	0.964	0.730	0.0184	0.0189	0.0303
750	----	61	0.111	0.998	0.782	0.0171	0.0176	0.0282
800	----	61	0.114	1.030	0.834	0.0161	0.0166	0.0265
900	----	61	0.122	1.094	0.940	0.0143	0.0147	0.0235
1000	----	61	0.128	1.152	1.042	0.0129	0.0132	0.0212
1250	----	91	0.117	1.289	1.305	0.0103	0.0106	0.0169
1500	----	91	0.128	1.412	1.566	0.00858	0.00883	0.0141
1750	----	127	0.117	1.526	1.829	0.00735	0.00756	0.0121
2000	----	127	0.126	1.632	2.092	0.00643	0.00662	0.0106

These resistance values are valid ONLY for the parameters as given. Using conductors having coated strands, different stranding type, and, especially, other temperatures changes the resistance.

Formula for temperature change: $R_2 = R_1 [1 + \alpha(T_2 - 75)]$ where $\alpha_{cu} = 0.00323$, $\alpha_{AL} = 0.00330$ at 75°C.

See NEC Chapter 9, Table 8. See Ugly's page 117 - 122 for metric conversions.

AC Resistance and Reactance for 600 Volt Cables, 3-Phase, 60 Hz, 75°C (167°F) – Three Single Conductors in Conduit

Ohms to Neutral per 1000 Feet															
Size	X_L (Reactance) for All Wires		AC Resistance for Uncoated Copper Wires			AC Resistance for Aluminum Wires			Effective Z at 0.85 PF for Uncoated Copper Wires			Effective Z at 0.85 PF for Aluminum Wires			Size
AWG/ kcmil	PVC, Al. Conduits	Steel Conduit	PVC Conduit	Al. Conduit	Steel Conduit	PVC Conduit	Al. Conduit	Steel Conduit	PVC Conduit	Al. Conduit	Steel Conduit	PVC Conduit	Al. Conduit	Steel Conduit	AWG/ kcmil
14	0.058	0.073	3.1	3.1	3.1	–	–	–	2.7	2.7	2.7	–	–	–	14
12	0.054	0.068	2.0	2.0	2.0	3.2	3.2	3.2	1.7	1.7	1.7	2.8	2.8	2.8	12
10	0.050	0.063	1.2	1.2	1.2	2.0	2.0	2.0	1.1	1.1	1.1	1.8	1.8	1.8	10
8	0.052	0.065	0.78	0.78	0.78	1.3	1.3	1.3	0.69	0.69	0.70	1.1	1.1	1.1	8
6	0.051	0.064	0.49	0.49	0.49	0.81	0.81	0.81	0.44	0.45	0.45	0.71	0.72	0.72	6
4	0.048	0.060	0.31	0.31	0.31	0.51	0.51	0.51	0.29	0.29	0.30	0.46	0.46	0.46	4
3	0.047	0.059	0.25	0.25	0.25	0.40	0.41	0.40	0.23	0.24	0.24	0.37	0.37	0.37	3
2	0.045	0.057	0.19	0.20	0.20	0.32	0.32	0.32	0.19	0.19	0.20	0.30	0.30	0.30	2
1	0.046	0.057	0.15	0.16	0.16	0.25	0.26	0.25	0.16	0.16	0.16	0.24	0.24	0.25	1
1/0	0.044	0.055	0.12	0.13	0.12	0.20	0.21	0.20	0.13	0.13	0.13	0.19	0.20	0.20	1/0
2/0	0.043	0.054	0.10	0.10	0.10	0.16	0.16	0.16	0.11	0.11	0.11	0.16	0.16	0.16	2/0
3/0	0.042	0.052	0.077	0.082	0.079	0.13	0.13	0.13	0.088	0.092	0.094	0.13	0.13	0.14	3/0
4/0	0.041	0.051	0.062	0.067	0.063	0.10	0.11	0.10	0.074	0.078	0.080	0.11	0.11	0.11	4/0
250	0.041	0.052	0.052	0.057	0.054	0.085	0.090	0.086	0.066	0.070	0.073	0.094	0.098	0.10	250
300	0.041	0.051	0.044	0.049	0.045	0.071	0.076	0.072	0.059	0.063	0.065	0.082	0.086	0.088	300
350	0.040	0.050	0.038	0.043	0.039	0.061	0.066	0.063	0.053	0.058	0.060	0.073	0.077	0.080	350
400	0.040	0.049	0.033	0.038	0.035	0.054	0.059	0.055	0.049	0.053	0.056	0.066	0.071	0.073	400
500	0.039	0.048	0.027	0.032	0.029	0.043	0.048	0.045	0.043	0.048	0.050	0.057	0.061	0.064	500
600	0.039	0.048	0.023	0.028	0.025	0.036	0.041	0.038	0.040	0.044	0.047	0.051	0.055	0.058	600
750	0.038	0.048	0.019	0.024	0.021	0.029	0.034	0.031	0.036	0.040	0.043	0.045	0.049	0.052	750
1000	0.037	0.046	0.015	0.019	0.018	0.023	0.027	0.025	0.032	0.036	0.040	0.039	0.042	0.046	1000

Notes: See NEC Table 9, page 70-568 for assumptions and explanations.
See Ugly's page 117 - 123 for metric conversion.

COPPER

AMPACITIES OF NOT MORE THAN **3** INSULATED CONDUCTORS RATED 0-2,000 VOLTS I**N RACEWAY OR CABLE**							
IN RACEWAY, CABLE, OR EARTH, BASED ON AMBIENT TEMPERATURE OF 30°C (86°F)				**IN RACEWAY OR CABLE**, BASED ON AMBIENT TEMPERATURE OF 40°C (104°F)			
SIZE	60°C (140°F)	75°C (167°F)	90°C (194°F)	150°C (302°F)	200°C (392°F)	250°C (482°F)	SIZE
AWG kcmil	**TYPES** TW, UF	**TYPES** RHW, THHW, THW, THWN, XHHW, USE, ZW	**TYPES** TBS, SA, SIS, FEP, FEPB, MI, RHH, RHW-2, THHN, THHW, THW-2,THWN-2, USE-2, XHH, XHHW, XHHW-2, ZW-2	**TYPES** Z,	**TYPES** FEP, FEPB, PFA	**TYPES** PFAH, TFE NICKEL OR NICKEL-COATED COPPER	AWG kcmil
14*	20	20	25	34	36	39	14
12*	25	25	30	43	45	54	12
10*	30	35	40	55	60	73	10
8	40	50	55	76	83	93	8
6	55	65	75	96	110	117	6
4	70	85	95	120	125	148	4
3	85	100	110	143	152	166	3
2	95	115	130	160	171	191	2
1	110	130	150	186	197	215	1
1/0	125	150	170	215	229	244	1/0
2/0	145	175	195	251	260	273	2/0
3/0	165	200	225	288	297	308	3/0
4/0	195	230	260	332	346	361	4/0
250	215	255	290	----	----	----	250
300	240	285	320	----	----	----	300
350	260	310	350	----	----	----	350
400	280	335	380	----	----	----	400
500	320	380	430	----	----	----	500
600	355	420	475	----	----	----	600
700	385	460	520	----	----	----	700
750	400	475	535	----	----	----	750
800	410	490	555	----	----	----	800
900	435	520	585	----	----	----	900
1000	455	545	615	----	----	----	1000
1250	495	590	665	----	----	----	1250
1500	520	625	705	----	----	----	1500
1750	545	650	735	----	----	----	1750
2000	560	665	750	----	----	----	2000

Overcurrent protection for conductor sizes marked (*) shall not exceed 15 amperes for No. 14, 20 amperes for No. 12, and 30 amperes for No. 10 copper. See NEC 240-2 for motor & welders overcurrent protection exceptions. See NEC 240.
For ambient temperatures other than those noted above, see NEC tables 310-16 & 310-18 Footnotes. See UGLY'S page 71 for adjustment factors for more than 3 conductors in raceway or cable.

ALUMINUM OR COPPER-CLAD ALUMINUM

AMPACITIES OF NOT MORE THAN **3** SINGLE INSULATED CONDUCTORS RATED 0 - 2,000 VOLTS **IN RACEWAY OR CABLE**					
	IN RACEWAY, CABLE OR EARTH BASED ON AMBIENT TEMPERATURE OF 30°C (86°F)			**IN RACEWAY OR CABLE** BASED ON AMBIENT TEMP OF 40°C (104°F)	
SIZE	60°C (140°F)	75°C (167°F)	90°C (194°F)	150°C (302°F)	SIZE
AWG kcmil	**TYPES** TW, UF	**TYPES** RHW, THHW, THW, THWN, XHHW, USE	**TYPES** TBS, SA, SIS, THHN, THHW, THW-2, THWN-2, RHH, RHW-2, USE-2, XHH, XHHW, XHHW-2, ZW-2	**TYPES** Z,	AWG kcmil
12*	20	20	25	30	12
10*	25	30	35	44	10
8	30	40	45	57	8
6	40	50	60	75	6
4	55	65	75	94	4
3	65	75	85	109	3
2	75	90	100	124	2
1	85	100	115	145	1
1/0	100	120	135	169	1/0
2/0	115	135	150	198	2/0
3/0	130	155	175	227	3/0
4/0	150	180	205	260	4/0
250	170	205	230	----	250
300	190	230	255	----	300
350	210	250	280	----	350
400	225	270	305	----	400
500	260	310	350	----	500
600	285	340	385	----	600
700	310	375	420	----	700
750	320	385	435	----	750
800	330	395	450	----	800
900	355	425	480	----	900
1000	375	445	500	----	1000
1250	405	485	545	----	1250
1500	435	520	585	----	1500
1750	455	545	615	----	1750
2000	470	560	630	----	2000

Overcurrent protection for conductor sizes marked (*) shall not exceed 15 amperes for Size 12 AWG, and 25 amperes for Size 10 AWG. See NEC 240-2 for motor & welders overcurrent protection exceptions. See NEC 240.
For ambient temperatures other than those noted above, see NEC tables 310-16 & 310-18 Footnotes. See UGLY'S page 71 for adjustment factors for more than 3 conductors in raceway or cable.

COPPER

AMPACITIES OF SINGLE INSULATED CONDUCTORS RATED 0-2,000 VOLTS **IN FREE AIR**							
	BASED ON AMBIENT TEMPERATURE OF 30°C (86°F)			BASED ON AMBIENT TEMPERATURE OF 40°C (104°F)			
SIZE	60°C (140°F)	75°C (167°F)	90°C (194°F)	150°C (302°F)	200°C (392°F)	250°C (482°F)	BARE OR COVERED CONDUC-TORS
AWG kcmil	**TYPES** TW, UF	**TYPES** RHW, THHW, THW, THWN, XHHW, ZW	**TYPES** TBS, SA, SIS, FEP, FEPB, MI, RHH, RHW-2, THHN, THHW, THW-2,THWN-2, USE-2, XHH, XHHW, XHHW-2, ZW-2	**TYPES** Z	**TYPES** FEP, FEPB, PFA	**TYPES** PFAH, TFE / NICKEL OR NICKEL-COATED COPPER	
14*	25	30	35	46	54	59	30
12*	30	35	40	60	68	78	35
10*	40	50	55	80	90	107	50
8	60	70	80	106	124	142	70
6	80	95	105	155	165	205	95
4	105	125	140	190	220	278	125
3	120	145	165	214	252	327	150
2	140	170	190	255	293	381	175
1	165	195	220	293	344	440	200
1/0	195	230	260	339	399	532	235
2/0	225	265	300	390	467	591	275
3/0	260	310	350	451	546	708	320
4/0	300	360	405	529	629	830	370
250	340	405	455	----	----	----	415
300	375	445	505	----	----	----	460
350	420	505	570	----	----	----	520
400	455	545	615	----	----	----	560
500	515	620	700	----	----	----	635
600	575	690	780	----	----	----	710
700	630	755	855	----	----	----	780
750	655	785	885	----	----	----	805
800	680	815	920	----	----	----	835
900	730	870	985	----	----	----	865
1000	780	935	1055	----	----	----	895
1250	890	1065	1200	----	----	----	----
1500	980	1175	1325	----	----	----	1205
1750	1070	1280	1445	----	----	----	----
2000	1155	1385	1560	----	----	----	1420

Overcurrent protection for conductor sizes marked (*) shall not exceed 15 amperes for No. 14, 20 amperes for No. 12, and 30 amperes for No. 10 copper. See NEC 240-3 for motor & welders overcurrent protection exceptions.
For ambient temperatures other than those noted above, see NEC tables 310-17 & 310-19.

ALUMINUM OR COPPER-CLAD ALUMINUM

AMPACITIES OF SINGLE INSULATED CONDUCTORS RATED 0-2,000 VOLTS **IN FREE AIR**					
	BASED ON AMBIENT TEMPERATURE OF 30°C (86°F)			BASED ON AMBIENT TEMP. OF 40°C (104°F)	
SIZE	60°C (140°F)	75°C (167°F)	90°C (194°F)	150°C (302°F)	SIZE
AWG kcmil	**TYPES** TW, UF	**TYPES** RHW, THHW, THW, THWN, XHHW	**TYPES** TBS, SA, SIS, RHH, RHW-2, THHN, THHW, THW-2, THWN-2, USE-2, XHH, ZW-2 XHHW, XHHW-2,	**TYPES** Z	AWG kcmil
14*	----	----	----	----	14
12*	25	30	35	47	12
10*	35	40	40	63	10
8	45	55	60	83	8
6	60	75	80	112	6
4	80	100	110	148	4
3	95	115	130	170	3
2	110	135	150	198	2
1	130	155	175	228	1
1/0	150	180	205	263	1/0
2/0	175	210	235	305	2/0
3/0	200	240	275	351	3/0
4/0	235	280	315	411	4/0
250	265	315	355	----	250
300	290	350	395	----	300
350	330	395	445	----	350
400	355	425	480	----	400
500	405	485	545	----	500
600	455	540	615	----	600
700	500	595	675	----	700
750	515	620	700	----	750
800	535	645	725	----	800
900	580	700	785	----	900
1000	625	750	845	----	1000
1250	710	855	960	----	1250
1500	795	950	1075	----	1500
1750	875	1050	1185	----	1750
2000	960	1150	1335	----	2000

Overcurrent protection for conductor sizes marked (*) shall not exceed 15 amperes for Size 12 AWG, and 25 amperes for Size 10 AWG. See NEC 240-3 for motor & welders overcurrent protection exceptions.
For ambient temperatures other than those noted above, see NEC tables 310-17 & 310-19.

ADJUSTMENT FACTORS

for More than Three Current-Carrying Conductors in a Raceway or Cable

Number of Current-Carrying Conductors	Percent of Values in Tables 310-16 through 310-19 as Adjusted for Ambient Temperature if Necessary
4 - 6	80
7 - 9	70
10 - 20	50
21 - 30	45
31 - 40	40
41 and above	35

INSULATION CHART

TRADE NAME	LETTER	MAX. TEMP.	APPLICATION PROVISIONS
FLOURINATED ETHYLENE PROPYLENE	FEP OR FEPB	90°C 194°F	DRY AND DAMP LOCATIONS
		200°C 392°F	DRY LOCATIONS - SPECIAL APPLICATIONS[2]
MINERAL INSULATION (METAL SHEATHED)	MI	90°C 194°F	DRY AND WET LOCATIONS
		250°C 482°F	SPECIAL APPLICATIONS[2]
MOISTURE, HEAT-, AND OIL-RESISTANT THERMOPLASTIC	MTW	60°C 140°F	MACHINE TOOL WIRING IN WET LOCATIONS, NFPA#79-ART. 670
		90°C 194°F	MACHINE TOOL WIRING IN DRY LOCATIONS, NFPA#79-ART. 670
PAPER		85°C 185°F	FOR UNDERGROUND SERVICE CONDUCTORS, OR BY SPECIAL PERMISSION
PERFLUOROALKOXY	PFA	90°C 194°F	DRY AND DAMP LOCATIONS
		200°C 392°F	DRY LOCATIONS - SPECIAL APPLICATIONS[2]

SEE *UGLY'S* PAGE 74 FOR SPECIAL PROVISIONS AND/OR APPLICATIONS.

INSULATION CHART

TRADE NAME	LETTER	MAX. TEMP.	APPLICATION PROVISIONS
PERFLUOROALKOXY	PFAH	250°C 482°F	DRY LOCATIONS ONLY. ONLY FOR LEADS WITHIN APPARATUS OR WITHIN RACEWAYS CONNECTED TO APPARATUS, (NICKEL OR NICKEL-COATED COPPER ONLY).
THERMOSET	RHH	90°C 194°F	DRY AND DAMP LOCATIONS
MOISTURE-RESISTANT THERMOSET	RHW[4]	75°C 167°F	DRY & WET LOCATIONS
MOISTURE-RESISTANT THERMOSET	RHW-2	90°C 194°F	DRY AND WET LOCATIONS
SILICONE	SA	90°C 194°F 200°C 392°F	DRY AND DAMP LOCATIONS FOR SPECIAL APPLICATIONS[2]
THERMOSET	SIS	90°C 194°F	SWITCHBOARD WIRING ONLY
THERMOPLASTIC AND FIBROUS OUTER BRAID	TBS	90°C 194°F	SWITCHBOARD WIRING ONLY
EXTENDED POLYTETRAFLUORO-ETHYLENE	TFE	250°C 482°F	DRY LOCATIONS ONLY. ONLY FOR LEADS WITHIN APPARATUS OR WITHIN RACEWAYS CONNECTED TO APPARATUS, OR AS OPEN WIRING (NICKEL OR NICKEL-COATED COPPER ONLY).
HEAT-RESISTANT THERMOPLASTIC	THHN	90°C 194°F	DRY AND DAMP LOCATIONS
MOISTURE-AND HEAT-RESISTANT THERMOPLASTIC	THHW	75°C 167°F 90°C 194°F	WET LOCATION DRY LOCATION

SEE *UGLY'S* PAGE 74 FOR SPECIAL PROVISIONS AND/OR APPLICATIONS.

INSULATION CHART

TRADE NAME	LETTER	MAX. TEMP.	APPLICATION PROVISIONS
MOISTURE - AND HEAT-RESISTANT THERMOPLASTIC	THW[4]	75°C 167°F 90°C 194°F	DRY & WET LOCATIONS SPECIAL APPL. WITHIN ELECTRIC DISCHARGE LIGHTING EQUIPMENT. LIMITED TO 1000 OPEN-CIRCUIT VOLTS OR LESS, (SIZE 14-8 ONLY AS PERMITTED IN SECTION 410-33)
MOISTURE - AND HEAT-RESISTANT THERMOPLASTIC	THWN[4]	75°C 167°F	DRY AND WET LOCATIONS
MOISTURE-RESISTANT THERMOPLASTIC	TW	60°C 140°F	DRY AND WET LOCATIONS
UNDERGROUND FEEDER AND BRANCH-CIRCUIT CABLE-SINGLE CONDUCTOR, (FOR TYPE "UF" CABLE EMPLOYING MORE THAN 1 CONDUCTOR. (SEE NEC ART. 340)	UF	60°C 140°F 75°C 167°F[7]	SEE ARTICLE 340 N.E.C.
UNDERGROUND SERVICE-ENTRANCE CABLE-SINGLE CONDUCTOR, (FOR TYPE "USE" CABLE EMPLOYING MORE THAN 1 CONDUCTOR. SEE N.E.C. ART. 338)	USE[4]	75°C 167°F	SEE ARTICLE 338 N.E.C.
THERMOSET	XHH	90°C 194°F	DRY AND DAMP LOCATIONS

SEE *UGLY'S* PAGE 74 FOR SPECIAL PROVISIONS AND/OR APPLICATIONS.

INSULATION CHART

TRADE NAME	LETTER	MAX. TEMP.	APPLICATION PROVISIONS
MOISTURE-RESISTANT THERMOSET	XHHW[4]	90°C 194°F 75°C 167°F	DRY AND DAMP LOCATIONS WET LOCATIONS
MOISTURE-RESISTANT THERMOSET	XHHW-2	90°C 194°F	DRY AND WET LOCATIONS
MODIFIED ETHYLENE TETRAFLUORO-ETHYLENE	Z	90°C 194°F 150°C 302°F	DRY AND DAMP LOCATIONS DRY LOCATIONS - SPECIAL APPLICATIONS[2]
MODIFIED ETHYLENE TETRAFLUORO-ETHYLENE	ZW[4]	75°C 167°F 90°C 194°F 150°C 302°F	WET LOCATIONS DRY AND DAMP LOCATIONS DRY LOCATIONS - SPECIAL APPLICATIONS[2]

FOOTNOTES:

1 Some insulations do not require an outer covering.
2 Where design conditions require maximum conductor operating temperatures above 90°C (194°F).
3 For signaling circuits permitting 300-volt insulation.
4 Listed wire types designated with the suffix "-2," such as RHW-2, shall be permitted to be used at a continuous 90°C (194°F) operating temperature, wet or dry.
5 Some rubber insulations do not require an outer covering.
6 Includes integral jacket.
7 For ampacity limitation, see Section 340.80 NEC.
8 Insulation thickness shall be permitted to be 2.03 mm (80 mils) for listed Type USE conductors that have been subjected to special investigations. The non-metallic covering over individual rubber-covered conductors of aluminum-sheathed cable and of lead-sheathed or multiconductor cable shall not be required to be flame retardant. For Type MC cable, see 330.104. For nonmetallic-sheathed cable, see Article 334, Part III. For Type UF cable, see Article 340, Part III.

MAXIMUM NUMBER OF CONDUCTORS IN TRADE SIZES OF CONDUIT OR TUBING

The 2002 National Electrical Code© shows 85 tables for conduit fill. There is a separate table for each type of conduit. In order to keep *Ugly's Electrical References©* in a compact and easy to use format, the following tables are included:

Electrical Metallic Tubing (EMT), Electrical Nonmetallic Tubing (ENT), PVC 40, PVC 80, Rigid Metal Conduit, Flexible Metal Conduit and Liquidtight Flexible Metal Conduit.

When other types of conduit are used, refer to Appendix C2002 *NEC* or use method shown below to figure conduit size.

Example #1 **- All same wire size and type insulation.**
10 – #12 THHW in Intermediate Metal Conduit.
Go to the THHW Conductor Square Inch Area Table. (Ugly's page 90)
#12 THHW = .026 sq. in.   10 x .026 sq. in. = .26 sq. in.
Go to Intermediate Metal Conduit Square Inch Area Table. (Ugly's page 93)
Use "Over 2 Wires 40%" column.
3/4 inch conduit = .235 sq. in. (less than .26, so it's too small).
1 inch conduit = .384 sq. in. (greater than .26, so it's correct size).

Example #2 **- Different wire sizes or types insulation.**
10 – #12 THHW and 10 – #10 THHN in Liquidtight Nonmetallic Conduit (FNMC-B).
Go to the THHW Conductor Square Inch Area Table. (Ugly's page 90)
#12 THHW = .026 sq. in.   10 x .026 sq. in. = .26 sq. in.
Go to the THHN Conductor Square Inch Area Table. (Ugly's page 90)
#10 THHN = .0211 sq. in.   10 x .0211 sq. in. = .211 sq. in.
.26 sq. in. + .211 sq. in. = .471 sq. in.
Go to Flexible Nonmetallic Conduit (FNMC-B) Square Inch Table. (Ugly's page 94)
Use "Over 2 Wires 40%" column.
1 inch conduit = .349 sq. in. (less than .471, so it's too small).
1 1/4 inch conduit = .611 sq. in. (greater than .471, so it's correct size).

NOTE 1:* All conductors must be counted including grounding conductors for fill percentage.
NOTE 2: When all conductors are same type and size, decimals .8 and larger must be rounded up.
*NOTE 3**:* These are minimum size calculations, under certain conditions jamming can occur and the next size conduit must be used.
*NOTE 4***:* CAUTION - When over three current carrying conductors are used in same circuit, conductor ampacity must be lower (derated).

* See Appendix C and Chapter Nine 2002 NEC for complete tables and examples.
** See Chapter nine Table one and Notes 1 - 10, 2002 NEC.
*** See notes to Ampacity Tables, Note 8, 2002 NEC.

MAXIMUM NUMBER OF CONDUCTORS IN ELECTRICAL METALLIC TUBING

Type Letters	Cond. Size AWG/kcmil	Trade Sizes In Inches									
		1/2	3/4	1	1 1/4	1 1/2	2	2 1/2	3	3 1/2	4
RHH, RHW, RHW-2	14	4	7	11	20	27	46	80	120	157	201
	12	3	6	9	17	23	38	66	100	131	167
	10	2	5	8	13	18	30	53	81	105	135
	8	1	2	4	7	9	16	28	42	55	70
	6	1	1	3	5	8	13	22	34	44	56
	4	1	1	2	4	6	10	17	26	34	44
	3	1	1	1	4	5	9	15	23	30	38
	2	1	1	1	3	4	7	13	20	26	33
	1	0	1	1	1	3	5	9	13	17	22
	1/0	0	1	1	1	2	4	7	11	15	19
	2/0	0	1	1	1	2	4	6	10	13	17
	3/0	0	0	1	1	1	3	5	8	11	14
	4/0	0	0	1	1	1	3	5	7	9	12
	250	0	0	0	1	1	1	3	5	7	9
	300	0	0	0	1	1	1	3	5	6	8
	350	0	0	0	1	1	1	3	4	6	7
	400	0	0	0	1	1	1	2	4	5	7
	500	0	0	0	0	1	1	2	3	4	6
	600	0	0	0	0	1	1	1	3	4	5
	700	0	0	0	0	0	1	1	2	3	4
	750	0	0	0	0	0	1	1	2	3	4
TW, THHW, THW, THW-2	14	8	15	25	43	58	96	168	254	332	424
	12	6	11	19	33	45	74	129	195	255	326
	10	5	8	14	24	33	55	96	145	190	243
	8	2	5	8	13	18	30	53	81	105	135
RHH*, RHW*, RHW-2*	14	6	10	16	28	39	64	112	169	221	282
	12	4	8	13	23	31	51	90	136	177	227
	10	3	6	10	18	24	40	70	106	138	177
	8	1	4	6	10	14	24	42	63	83	106
RHH*, RHW*, RHW-2*, TW, THW, THHW, THW-2	6	1	3	4	8	11	18	32	48	63	81
	4	1	1	3	6	8	13	24	36	47	60
	3	1	1	3	5	7	12	20	31	40	52
	2	1	1	2	4	6	10	17	26	34	44
	1	1	1	1	3	4	7	12	18	24	31
	1/0	0	1	1	2	3	6	10	16	20	26
	2/0	0	1	1	1	3	5	9	13	17	22
	3/0	0	1	1	1	2	4	7	11	15	19
	4/0	0	0	1	1	1	3	6	9	12	16
	250	0	0	1	1	1	3	5	7	10	13
	300	0	0	1	1	1	2	4	6	8	11
	350	0	0	0	1	1	1	4	6	7	10
	400	0	0	0	1	1	1	3	5	7	9
	500	0	0	0	1	1	1	3	4	6	7
	600	0	0	0	1	1	1	2	3	4	6
	700	0	0	0	0	1	1	1	3	4	5
	750	0	0	0	0	1	1	1	3	4	5
THHN, THWN, THWN-2	14	12	22	35	61	84	138	241	364	476	608
	12	9	16	26	45	61	101	176	266	347	443
	10	5	10	16	28	38	63	111	167	219	279
	8	3	6	9	16	22	36	64	96	126	161
	6	2	4	7	12	16	26	46	69	91	116
	4	1	2	4	7	10	16	28	43	56	71
	3	1	1	3	6	8	13	24	36	47	60
	2	1	1	3	5	7	11	20	30	40	51
	1	1	1	1	4	5	8	15	22	29	37
	1/0	1	1	1	3	4	7	12	19	25	32
	2/0	0	1	1	2	3	6	10	16	20	26
	3/0	0	1	1	1	3	5	8	13	17	22

MAXIMUM NUMBER OF CONDUCTORS IN *ELECTRICAL METALLIC TUBING*

Type Letters	Cond. Size AWG/kcmil	Trade Sizes In Inches 1/2	3/4	1	1 1/4	1 1/2	2	2 1/2	3	3 1/2	4
THHN, THWN, THWN-2	4/0	0	1	1	1	2	4	7	11	14	18
	250	0	0	1	1	1	3	6	9	11	15
	300	0	0	1	1	1	3	5	7	10	13
	350	0	0	1	1	1	2	4	6	9	11
	400	0	0	0	1	1	1	4	6	8	10
	500	0	0	0	1	1	1	3	5	6	8
	600	0	0	0	1	1	1	2	4	5	7
	700	0	0	0	1	1	1	2	3	4	6
	750	0	0	0	0	1	1	1	3	4	5
FEP, FEPB, PFA, PFAH, TFE	14	12	21	34	60	81	134	234	354	462	590
	12	9	15	25	43	59	98	171	258	337	430
	10	6	11	18	31	42	70	122	185	241	309
	8	3	6	10	18	24	40	70	106	138	177
	6	2	4	7	12	17	28	50	75	98	126
	4	1	3	5	9	12	20	35	53	69	88
	3	1	2	4	7	10	16	29	44	57	73
	2	1	1	3	6	8	13	24	36	47	60
PFA, PFAH, TFE	1	1	1	2	4	6	9	16	25	33	42
PFA, PFAH, TFE, Z	1/0	1	1	1	3	5	8	14	21	27	35
	2/0	0	1	1	3	4	6	11	17	22	29
	3/0	0	1	1	2	3	5	9	14	18	24
	4/0	0	1	1	1	2	4	8	11	15	19
Z	14	14	25	41	72	98	161	282	426	556	711
	12	10	18	29	51	69	114	200	302	394	504
	10	6	11	18	31	42	70	122	185	241	309
	8	4	7	11	20	27	44	77	117	153	195
	6	3	5	8	14	19	31	54	82	107	137
	4	1	3	5	9	13	21	37	56	74	94
	3	1	2	4	7	9	15	27	41	54	69
	2	1	1	3	6	8	13	22	34	45	57
	1	1	1	2	4	6	10	18	28	36	46
XHH, XHHW, XHHW-2, ZW	14	8	15	25	43	58	96	168	254	332	424
	12	6	11	19	33	45	74	129	195	255	326
	10	5	8	14	24	33	55	96	145	190	243
	8	2	5	8	13	18	30	53	81	105	135
	6	1	3	6	10	14	22	39	60	78	100
	4	1	2	4	7	10	16	28	43	56	72
	3	1	1	3	6	8	14	24	36	48	61
	2	1	1	3	5	7	11	20	31	40	51
XHH, XHHW, XHHW-2	1	1	1	1	4	5	8	15	23	30	38
	1/0	1	1	1	3	4	7	13	19	25	32
	2/0	0	1	1	2	3	6	10	16	21	27
	3/0	0	1	1	1	3	5	9	13	17	22
	4/0	0	1	1	1	2	4	7	11	14	18
	250	0	0	1	1	1	3	6	9	12	15
	300	0	0	1	1	1	3	5	8	10	13
	350	0	0	1	1	1	2	4	7	9	11
	400	0	0	0	1	1	1	4	6	8	10
	500	0	0	0	1	1	1	3	5	6	8
	600	0	0	0	1	1	1	2	4	5	6
	700	0	0	0	0	1	1	2	3	4	6
	750	0	0	0	0	1	1	1	3	4	5

* Types RHH, RHW, AND RHW-2 without outer covering.

See Ugly's page 119 for Trade Size / Metric Designator conversion.

MAXIMUM NUMBER OF CONDUCTORS IN NONMETALLIC TUBING

Type Letters	Cond. Size AWG/kcmil	Trade Sizes In Inches ½	¾	1	1¼	1½	2
RHH, RHW, RHW-2	14	3	6	10	19	26	43
	12	2	5	9	16	22	36
	10	1	4	7	13	17	29
	8	1	1	3	6	9	15
	6	1	1	3	5	7	12
	4	1	1	2	4	6	9
	3	1	1	1	3	5	8
	2	0	1	1	3	4	7
	1	0	1	1	1	3	5
	1/0	0	0	1	1	2	4
	2/0	0	0	1	1	1	3
	3/0	0	0	1	1	1	3
	4/0	0	0	1	1	1	2
	250	0	0	0	1	1	1
	300	0	0	0	1	1	1
	350	0	0	0	1	1	1
	400	0	0	0	1	1	1
	500	0	0	0	0	1	1
	600	0	0	0	0	1	1
	700	0	0	0	0	0	1
	750	0	0	0	0	0	1
TW THHW, THW, THW-2	14	7	13	22	40	55	92
	12	5	10	17	31	42	71
	10	4	7	13	23	32	52
	8	1	4	7	13	17	29
RHH*, RHW*, RHW-2*	14	4	8	15	27	37	61
	12	3	7	12	21	29	49
	10	3	5	9	17	23	38
	8	1	3	5	10	14	23
RHH*, RHW*, RHW-2*, TW, THW, THHW, THW-2	6	1	2	4	7	10	17
	4	1	1	3	5	8	13
	3	1	1	2	5	7	11
	2	1	1	2	4	6	9
	1	0	1	1	3	4	6
	1/0	0	1	1	2	3	5
	2/0	0	1	1	1	3	5
	3/0	0	0	1	1	2	4
	4/0	0	0	1	1	1	3
	250	0	0	1	1	1	2
	300	0	0	0	1	1	2
	350	0	0	0	1	1	1
	400	0	0	0	1	1	1
	500	0	0	0	1	1	1
	600	0	0	0	0	1	1
	700	0	0	0	0	1	1
	750	0	0	0	0	1	1
THHN, THWN, THWN-2	14	10	18	32	58	80	132
	12	7	13	23	42	58	96
	10	4	8	15	26	36	60
	8	2	5	8	15	21	35
	6	1	3	6	11	15	25
	4	1	1	4	7	9	15
	3	1	1	3	5	8	13
	2	1	1	2	5	6	11
	1	1	1	1	3	5	8
	1/0	0	1	1	3	4	7
	2/0	0	1	1	2	3	5
	3/0	0	1	1	1	3	4

MAXIMUM NUMBER OF CONDUCTORS IN *NONMETALLIC TUBING*

Type Letters	Cond. Size AWG/kcmil	1/2	3/4	1	1 1/4	1 1/2	2
THHN, THWN, THWN-2	4/0	0	0	1	1	2	4
	250	0	0	1	1	1	3
	300	0	0	1	1	1	2
	350	0	0	0	1	1	2
	400	0	0	0	1	1	1
	500	0	0	0	1	1	1
	600	0	0	0	1	1	1
	700	0	0	0	0	1	1
	750	0	0	0	0	1	1
FEP, FEPB, PFA, PFAH, TFE	14	10	18	31	56	77	128
	12	7	13	23	41	56	93
	10	5	9	16	29	40	67
	8	3	5	9	17	23	38
	6	1	4	6	12	16	27
	4	1	2	4	8	11	19
	3	1	1	4	7	9	16
	2	1	1	3	5	8	13
PFA, PFAH, TFE	1	1	1	1	4	5	9
PFA, PFAH, TFE, Z	1/0	0	1	1	3	4	7
	2/0	0	1	1	2	4	6
	3/0	0	1	1	1	3	5
	4/0	0	1	1	1	2	4
Z	14	12	22	38	68	93	154
	12	8	15	27	48	66	109
	10	5	9	16	29	40	67
	8	3	6	10	18	25	42
	6	1	4	7	13	18	30
	4	1	3	5	9	12	20
	3	1	1	3	6	9	15
	2	1	1	3	5	7	12
	1	1	1	2	4	6	10
XHH, XHHW, XHHW-2, ZW	14	7	13	22	40	55	92
	12	5	10	17	31	42	71
	10	4	7	13	23	32	52
	8	1	4	7	13	17	29
	6	1	3	5	9	13	21
	4	1	1	4	7	9	15
	3	1	1	3	6	8	13
	2	1	1	2	5	6	11
XHH, XHHW, XHHW-2	1	1	1	1	3	5	8
	1/0	0	1	1	3	4	7
	2/0	0	1	1	2	3	6
	3/0	0	1	1	1	3	5
	4/0	0	0	1	1	2	4
	250	0	0	1	1	1	3
	300	0	0	1	1	1	3
	350	0	0	1	1	1	2
	400	0	0	0	1	1	1
	500	0	0	0	1	1	1
	600	0	0	0	1	1	1
	700	0	0	0	0	1	1
	750	0	0	0	0	1	1

* Types RHH, RHW, AND RHW-2 without outer covering.

See Ugly's page 119 for Trade Size / Metric Designator conversion.

MAXIMUM NUMBER OF CONDUCTORS IN RIGID PVC CONDUIT, SCHEDULE 40

Type Letters	Cond. Size AWG/kcmil	Trade Sizes In Inches 1/2	3/4	1	1 1/4	1 1/2	2	2 1/2	3	3 1/2	4	5	6
RHH, RHW, RHW-2	14	4	7	11	20	27	45	64	99	133	171	269	390
	12	3	5	9	16	22	37	53	82	110	142	224	323
	10	2	4	7	13	18	30	43	66	89	115	181	261
	8	1	2	4	7	9	15	22	35	46	60	94	137
	6	1	1	3	5	7	12	18	28	37	48	76	109
	4	1	1	2	4	6	10	14	22	29	37	59	85
	3	1	1	1	4	5	8	12	19	25	33	52	75
	2	1	1	1	3	4	7	10	16	22	28	45	65
	1	0	1	1	1	3	5	7	11	14	19	29	43
	1/0	0	1	1	1	2	4	6	9	13	16	26	37
	2/0	0	0	1	1	1	3	5	8	11	14	22	32
	3/0	0	0	1	1	1	3	4	7	9	12	19	28
	4/0	0	0	1	1	1	2	4	6	8	10	16	24
	250	0	0	0	1	1	1	3	4	6	8	12	18
	300	0	0	0	1	1	1	2	4	5	7	11	16
	350	0	0	0	1	1	1	2	3	5	6	10	14
	400	0	0	0	1	1	1	1	3	4	6	9	13
	500	0	0	0	0	1	1	1	3	4	5	8	11
	600	0	0	0	0	1	1	1	2	3	4	6	9
	700	0	0	0	0	0	1	1	1	3	3	6	8
	750	0	0	0	0	0	1	1	1	2	3	5	8
TW THHW, THW, THW-2	14	8	14	24	42	57	94	135	209	280	361	568	822
	12	6	11	18	32	44	72	103	160	215	277	436	631
	10	4	8	13	24	32	54	77	119	160	206	325	470
	8	2	4	7	13	18	30	43	66	89	115	181	261
RHH*, RHW*, RHW-2*	14	5	9	16	28	38	63	90	139	186	240	378	546
	12	4	8	12	22	30	50	72	112	150	193	304	439
	10	3	6	10	17	24	39	56	87	117	150	237	343
	8	1	3	6	10	14	23	33	52	70	90	142	205
TW, THW, THHW, THW-2	6	1	2	4	8	11	18	26	40	53	69	109	157
	4	1	1	3	6	8	13	19	30	40	51	81	117
	3	1	1	3	5	7	11	16	25	34	44	69	100
	2	1	1	2	4	6	10	14	22	29	37	59	85
	1	0	1	1	3	4	7	10	15	20	26	41	60
	1/0	0	1	1	2	3	6	8	13	17	22	35	51
	2/0	0	1	1	1	3	5	7	11	15	19	30	43
	3/0	0	1	1	1	2	4	6	9	12	16	25	36
	4/0	0	0	1	1	1	3	5	8	10	13	21	30
	250	0	0	1	1	1	3	4	6	8	11	17	25
	300	0	0	1	1	1	2	3	5	7	9	15	21
	350	0	0	0	1	1	1	3	5	6	8	13	19
	400	0	0	0	1	1	1	3	4	6	7	12	17
	500	0	0	0	1	1	1	2	3	5	6	10	14
	600	0	0	0	0	1	1	1	3	4	5	8	11
	700	0	0	0	0	1	1	1	2	3	4	7	10
	750	0	0	0	0	1	1	1	2	3	4	6	10
THHN, THWN, THWN-2	14	11	21	34	60	82	135	193	299	401	517	815	1178
	12	8	15	25	43	59	99	141	218	293	377	594	859
	10	5	9	15	27	37	62	89	137	184	238	374	541
	8	3	5	9	16	21	36	51	79	106	137	216	312
	6	1	4	6	11	15	26	37	57	77	99	156	225
	4	1	2	4	7	9	16	22	35	47	61	96	138
	3	1	1	3	6	8	13	19	30	40	51	81	117
	2	1	1	3	5	7	11	16	25	33	43	68	98
	1	1	1	1	3	5	8	12	18	25	32	50	73
	1/0	1	1	1	3	4	7	10	15	21	27	42	61
	2/0	0	1	1	2	3	6	8	13	17	22	35	51
	3/0	0	1	1	1	3	5	7	11	14	18	29	42

MAXIMUM NUMBER OF CONDUCTORS IN RIGID PVC CONDUIT, SCHEDULE 40

Type Letters	Cond. Size AWG/kcmil	Trade Sizes In Inches ½	¾	1	1¼	1½	2	2½	3	3½	4	5	6
THHN, THWN, THWN-2	4/0	0	1	1	1	2	4	6	9	12	15	24	35
	250	0	0	1	1	1	3	4	7	10	12	20	28
	300	0	0	1	1	1	3	4	6	8	11	17	24
	350	0	0	1	1	1	2	3	5	7	9	15	21
	400	0	0	0	1	1	1	3	5	6	8	13	19
	500	0	0	0	1	1	1	2	4	5	7	11	16
	600	0	0	0	1	1	1	1	3	4	5	9	13
	700	0	0	0	0	1	1	1	3	4	5	8	11
	750	0	0	0	0	1	1	1	2	3	4	7	11
FEP, FEPB, PFA, PFAH, TFE	14	11	20	33	58	79	131	188	290	389	502	790	1142
	12	8	15	24	42	58	96	137	212	284	366	577	834
	10	6	10	17	30	41	69	98	152	204	263	414	598
	8	3	6	10	17	24	39	56	87	117	150	237	343
	6	2	4	7	12	17	28	40	62	83	107	169	244
	4	1	3	5	8	12	19	28	43	58	75	118	170
	3	1	2	4	7	10	16	23	36	48	62	98	142
	2	1	1	3	6	8	13	19	30	40	51	81	117
PFA, PFAH, TFE	1	1	1	2	4	5	9	13	20	28	36	56	81
PFA, PFAH, TFE, Z	1/0	1	1	1	3	4	8	11	17	23	30	47	68
	2/0	0	1	1	3	4	6	9	14	19	24	39	56
	3/0	0	1	1	2	3	5	7	12	16	20	32	46
	4/0	0	1	1	1	2	4	6	9	13	16	26	38
Z	14	13	24	40	70	95	158	226	350	469	605	952	1376
	12	9	17	28	49	68	112	160	248	333	429	675	976
	10	6	10	17	30	41	69	98	152	204	263	414	598
	8	3	6	11	19	26	43	62	96	129	166	261	378
	6	2	4	7	13	18	30	43	67	90	116	184	265
	4	1	3	5	9	12	21	30	46	62	80	126	183
	3	1	2	4	6	9	15	22	34	45	58	92	133
	2	1	1	3	5	7	12	18	28	38	49	77	111
	1	1	1	2	4	6	10	14	23	30	39	62	90
XHH, XHHW, XHHW-2, ZW	14	8	14	24	42	57	94	135	209	280	361	568	822
	12	6	11	18	32	44	72	103	160	215	277	436	631
	10	4	8	13	24	32	54	77	119	160	206	325	470
	8	2	4	7	13	18	30	43	66	89	115	181	261
	6	1	3	5	10	13	22	32	49	66	85	134	193
	4	1	2	4	7	9	16	23	35	48	61	97	140
	3	1	1	3	6	8	13	19	30	40	52	82	118
	2	1	1	3	5	7	11	16	25	34	44	69	99
XHH, XHHW, XHHW-2	1	1	1	1	3	5	8	12	19	25	32	51	74
	1/0	1	1	1	3	4	7	10	16	21	27	43	62
	2/0	0	1	1	2	3	6	8	13	17	23	36	52
	3/0	0	1	1	1	3	5	7	11	14	19	30	43
	4/0	0	1	1	1	2	4	6	9	12	15	24	35
	250	0	0	1	1	1	3	5	7	10	13	20	29
	300	0	0	1	1	1	3	4	6	8	11	17	25
	350	0	0	1	1	1	2	3	5	7	9	15	22
	400	0	0	0	1	1	1	3	5	6	8	13	19
	500	0	0	0	1	1	1	2	4	5	7	11	16
	600	0	0	0	1	1	1	1	3	4	5	9	13
	700	0	0	0	0	1	1	1	3	4	5	8	11
	750	0	0	0	0	1	1	1	2	3	4	7	11

* Types RHH, RHW, AND RHW-2 without outer covering.

See Ugly's page 119 for Trade Size / Metric Designator conversion.

MAXIMUM NUMBER OF CONDUCTORS IN *RIGID PVC CONDUIT, SCHEDULE 80*

Type Letters	Cond. Size AWG/kcmil	Trade Sizes In Inches ½	¾	1	1¼	1½	2	2½	3	3½	4	5	6
RHH, RHW, RHW-2	14	3	5	9	17	23	39	56	88	118	153	243	349
	12	2	4	7	14	19	32	46	73	98	127	202	290
	10	1	3	6	11	15	26	37	59	79	103	163	234
	8	1	1	3	6	8	13	19	31	41	54	85	122
	6	1	1	2	4	6	11	16	24	33	43	68	98
	4	1	1	1	3	5	8	12	19	26	33	53	77
	3	0	1	1	3	4	7	11	17	23	29	47	67
	2	0	1	1	3	4	6	9	14	20	25	41	58
	1	0	1	1	1	2	4	6	9	13	17	27	38
	1/0	0	0	1	1	1	3	5	8	11	15	23	33
	2/0	0	0	1	1	1	3	4	7	10	13	20	29
	3/0	0	0	1	1	1	3	4	6	8	11	17	25
	4/0	0	0	0	1	1	2	3	5	7	9	15	21
	250	0	0	0	1	1	1	2	4	5	7	11	16
	300	0	0	0	1	1	1	2	3	5	6	10	14
	350	0	0	0	1	1	1	1	3	4	5	9	13
	400	0	0	0	0	1	1	1	3	4	5	8	12
	500	0	0	0	0	1	1	1	2	3	4	7	10
	600	0	0	0	0	0	1	1	1	3	3	6	8
	700	0	0	0	0	0	1	1	1	2	3	5	7
	750	0	0	0	0	0	1	1	1	2	3	5	7
TW THHW, THW, THW-2	14	6	11	20	35	49	82	118	185	250	324	514	736
	12	5	9	15	27	38	63	91	142	192	248	394	565
	10	3	6	11	20	28	47	67	106	143	185	294	421
	8	1	3	6	11	15	26	37	59	79	103	163	234
RHH*, RHW*, RHW-2*	14	4	8	13	23	32	55	79	123	166	215	341	490
	12	3	6	10	19	26	44	63	99	133	173	274	394
	10	2	5	8	15	20	34	49	77	104	135	214	307
	8	1	3	5	9	12	20	29	46	62	81	128	184
RHH*, RHW*, RHW-2*, TW, THW, THHW, THW-2	6	1	1	3	7	9	16	22	35	48	62	98	141
	4	1	1	3	5	7	12	17	26	35	46	73	105
	3	1	1	2	4	6	10	14	22	30	39	63	90
	2	1	1	1	3	5	8	12	19	26	33	53	77
	1	0	1	1	2	3	6	8	13	18	23	37	54
	1/0	0	1	1	1	3	5	7	11	15	20	32	46
	2/0	0	1	1	1	2	4	6	10	13	17	27	39
	3/0	0	0	1	1	1	3	5	8	11	14	23	33
	4/0	0	0	1	1	1	3	4	7	9	12	19	27
	250	0	0	0	1	1	2	3	5	7	9	15	22
	300	0	0	0	1	1	1	3	5	6	8	13	19
	350	0	0	0	1	1	1	2	4	6	7	12	17
	400	0	0	0	1	1	1	2	4	5	7	10	15
	500	0	0	0	1	1	1	1	3	4	5	9	13
	600	0	0	0	0	1	1	1	2	3	4	7	10
	700	0	0	0	0	1	1	1	2	3	4	6	9
	750	0	0	0	0	0	1	1	1	3	4	6	8
THHN, THWN, THWN-2	14	9	17	28	51	70	118	170	265	358	464	736	1055
	12	6	12	20	37	51	86	124	193	261	338	537	770
	10	4	7	13	23	32	54	78	122	164	213	338	485
	8	2	4	7	13	18	31	45	70	95	123	195	279
	6	1	3	5	9	13	22	32	51	68	89	141	202
	4	1	1	3	6	8	14	20	31	42	54	86	124
	3	1	1	3	5	7	12	17	26	35	46	73	105
	2	1	1	2	4	6	10	14	22	30	39	61	88
	1	0	1	1	3	4	7	10	16	22	29	45	65
	1/0	0	1	1	2	3	6	9	14	18	24	38	55
	2/0	0	1	1	1	3	5	7	11	15	20	32	46
	3/0	0	1	1	1	2	4	6	9	13	17	26	38

MAXIMUM NUMBER OF CONDUCTORS IN *RIGID PVC CONDUIT, SCHEDULE 80*

Type Letters	Cond. Size AWG/kcmil	Trade Sizes In Inches ½	¾	1	1¼	1½	2	2½	3	3½	4	5	6
THHN, THWN, THWN-2	4/0	0	0	1	1	1	3	5	8	10	14	22	31
	250	0	0	1	1	1	3	4	6	8	11	18	25
	300	0	0	0	1	1	2	3	5	7	9	15	22
	350	0	0	0	1	1	1	3	5	6	8	13	19
	400	0	0	0	1	1	1	3	4	6	7	12	17
	500	0	0	0	1	1	1	2	3	5	6	10	14
	600	0	0	0	0	1	1	1	3	4	5	8	12
	700	0	0	0	0	1	1	1	2	3	4	7	10
	750	0	0	0	0	1	1	1	2	3	4	7	9
FEP, FEPB, PFA, PFAH, TFE	14	8	16	27	49	68	115	164	257	347	450	714	1024
	12	6	12	20	36	50	84	120	188	253	328	521	747
	10	4	8	14	26	36	60	86	135	182	235	374	536
	8	2	5	8	15	20	34	49	77	104	135	214	307
	6	1	3	6	10	14	24	35	55	74	96	152	218
	4	1	2	4	7	10	17	24	38	52	67	106	153
	3	1	1	3	6	8	14	20	32	43	56	89	127
	2	1	1	3	5	7	12	17	26	35	46	73	105
PFA, PFAH, TFE	1	1	1	1	3	5	8	11	18	25	32	51	73
PFA, PFAH, TFE, Z	1/0	0	1	1	3	4	7	10	15	20	27	42	61
	2/0	0	1	1	2	3	5	8	12	17	22	35	50
	3/0	0	1	1	1	2	4	6	10	14	18	29	41
	4/0	0	0	1	1	1	4	5	8	11	15	24	34
Z	14	10	19	33	59	82	138	198	310	418	542	860	1233
	12	7	14	23	42	58	98	141	220	297	385	610	875
	10	4	8	14	26	36	60	86	135	182	235	374	536
	8	3	5	9	16	22	38	54	85	115	149	236	339
	6	2	4	6	11	16	26	38	60	81	104	166	238
	4	1	2	4	8	11	18	26	41	55	72	114	164
	3	1	2	3	5	8	13	19	30	40	52	83	119
	2	1	1	2	5	6	11	16	25	33	43	69	99
	1	0	1	2	4	5	9	13	20	27	35	56	80
XHH, XHHW, XHHW-2, ZW	14	6	11	20	35	49	82	118	185	250	324	514	736
	12	5	9	15	27	38	63	91	142	192	248	394	565
	10	3	6	11	20	28	47	67	106	143	185	294	421
	8	1	3	6	11	15	26	37	59	79	103	163	234
	6	1	2	4	8	11	19	28	43	59	76	121	173
	4	1	1	3	6	8	14	20	31	42	55	87	125
	3	1	1	3	5	7	12	17	26	36	47	74	106
	2	1	1	2	4	6	10	14	22	30	39	62	89
XHH, XHHW, XHHW-2	1	0	1	1	3	4	7	10	16	22	29	46	66
	1/0	0	1	1	2	3	6	9	14	19	24	39	56
	2/0	0	1	1	1	3	5	7	11	16	20	32	46
	3/0	0	1	1	1	2	4	6	9	13	17	27	38
	4/0	0	0	1	1	1	3	5	8	11	14	22	32
	250	0	0	1	1	1	3	4	6	9	11	18	26
	300	0	0	1	1	1	2	3	5	7	10	15	22
	350	0	0	0	1	1	1	3	5	6	8	14	20
	400	0	0	0	1	1	1	3	4	6	7	12	17
	500	0	0	0	1	1	1	2	3	5	6	10	14
	600	0	0	0	0	1	1	1	3	4	5	8	11
	700	0	0	0	0	1	1	1	2	3	4	7	10
	750	0	0	0	0	1	1	1	2	3	4	6	9

* Types RHH, RHW, AND RHW-2 without outer covering.

See Ugly's page 119 for Trade Size / Metric Designator conversion.

MAXIMUM NUMBER OF CONDUCTORS IN RIGID METAL CONDUIT

Type Letters	Cond. Size AWG/kcmil	Trade Sizes In Inches ½	¾	1	1¼	1½	2	2½	3	3½	4	5	6
RHH, RHW, RHW-2	14	4	7	12	21	28	46	66	102	136	176	276	398
	12	3	6	10	17	23	38	55	85	113	146	229	330
	10	3	5	8	14	19	31	44	68	91	118	185	267
	8	1	2	4	7	10	16	23	36	48	61	97	139
	6	1	1	3	6	8	13	18	29	38	49	77	112
	4	1	1	2	4	6	10	14	22	30	38	60	87
	3	1	1	2	4	5	9	12	19	26	34	53	76
	2	1	1	1	3	4	7	11	17	23	29	46	66
	1	0	1	1	1	3	5	7	11	15	19	30	44
	1/0	0	1	1	1	2	4	6	10	13	17	26	38
	2/0	0	1	1	1	2	4	5	8	11	14	23	33
	3/0	0	0	1	1	1	3	4	7	10	12	20	28
	4/0	0	0	1	1	1	3	4	6	8	11	17	24
	250	0	0	0	1	1	1	3	4	6	8	13	18
	300	0	0	0	1	1	1	2	4	5	7	11	16
	350	0	0	0	1	1	1	2	4	5	6	10	15
	400	0	0	0	1	1	1	1	3	4	6	9	13
	500	0	0	0	1	1	1	1	3	4	5	8	11
	600	0	0	0	0	1	1	1	2	3	4	6	9
	700	0	0	0	0	1	1	1	1	3	4	6	8
	750	0	0	0	0	0	1	1	1	3	3	5	8
TW THHW, THW, THW-2	14	9	15	25	44	59	98	140	216	288	370	581	839
	12	7	12	19	33	45	75	107	165	221	284	446	644
	10	5	9	14	25	34	56	80	123	164	212	332	480
	8	3	5	8	14	19	31	44	68	91	118	185	267
RHH*, RHW*, RHW-2*	14	6	10	17	29	39	65	93	143	191	246	387	558
	12	5	8	13	23	32	52	75	115	154	198	311	448
	10	3	6	10	18	25	41	58	90	120	154	242	350
	8	1	4	6	11	15	24	35	54	72	92	145	209
RHH*, RHW*, RHW-2*, TW, THW, THHW, THW-2	6	1	3	5	8	11	18	27	41	55	71	111	160
	4	1	1	3	6	8	14	20	31	41	53	83	120
	3	1	1	3	5	7	12	17	26	35	45	71	103
	2	1	1	2	4	6	10	14	22	30	38	60	87
	1	1	1	1	3	4	7	10	15	21	27	42	61
	1/0	0	1	1	2	3	6	8	13	18	23	36	52
	2/0	0	1	1	2	3	5	7	11	15	19	31	44
	3/0	0	1	1	1	2	4	6	9	13	16	26	37
	4/0	0	0	1	1	1	3	5	8	10	14	21	31
	250	0	0	1	1	1	3	4	6	8	11	17	25
	300	0	0	1	1	1	2	3	5	7	9	15	22
	350	0	0	0	1	1	1	3	5	6	8	13	19
	400	0	0	0	1	1	1	3	4	6	7	12	17
	500	0	0	0	1	1	1	2	3	5	6	10	14
	600	0	0	0	1	1	1	1	3	4	5	8	12
	700	0	0	0	0	1	1	1	2	3	4	7	10
	750	0	0	0	0	1	1	1	2	3	4	7	10
THHN, THWN, THWN-2	14	13	22	36	63	85	140	200	309	412	531	833	1202
	12	9	16	26	46	62	102	146	225	301	387	608	877
	10	6	10	17	29	39	64	92	142	189	244	383	552
	8	3	6	9	16	22	37	53	82	109	140	221	318
	6	2	4	7	12	16	27	38	59	79	101	159	230
	4	1	2	4	7	10	16	23	36	48	62	98	141
	3	1	1	3	6	8	14	20	31	41	53	83	120
	2	1	1	3	5	7	11	17	26	34	44	70	100
	1	1	1	1	4	5	8	12	19	25	33	51	74
	1/0	1	1	1	3	4	7	10	16	21	27	43	63
	2/0	0	1	1	2	3	6	8	13	18	23	36	52
	3/0	0	1	1	1	3	5	7	11	15	19	30	43

MAXIMUM NUMBER OF CONDUCTORS IN *RIGID METAL CONDUIT*

Type Letters	Cond. Size AWG/kcmil	Trade Sizes In Inches 1/2	3/4	1	1 1/4	1 1/2	2	2 1/2	3	3 1/2	4	5	6
THHN, THWN, THWN-2	4/0	0	1	1	1	2	4	6	9	12	16	25	36
	250	0	0	1	1	1	3	5	7	10	13	20	29
	300	0	0	1	1	1	3	4	6	8	11	17	25
	350	0	0	1	1	1	2	3	5	7	10	15	22
	400	0	0	1	1	1	2	3	5	7	8	13	20
	500	0	0	0	1	1	1	2	4	5	7	11	16
	600	0	0	0	1	1	1	1	3	4	6	9	13
	700	0	0	0	1	1	1	1	3	4	5	8	11
	750	0	0	0	0	1	1	1	3	4	5	7	11
FEP, FEPB, PFA, PFAH, TFE	14	12	22	35	61	83	136	194	300	400	515	808	1166
	12	9	16	26	44	60	99	142	219	292	376	590	851
	10	6	11	18	32	43	71	102	157	209	269	423	610
	8	3	6	10	18	25	41	58	90	120	154	242	350
	6	2	4	7	13	17	29	41	64	85	110	172	249
	4	1	3	5	9	12	20	29	44	59	77	120	174
	3	1	2	4	7	10	17	24	37	50	64	100	145
	2	1	1	3	6	8	14	20	31	41	53	83	120
PFA, PFAH, TFE	1	1	1	2	4	6	9	14	21	28	37	57	83
PFA, PFAH, TFE, Z	1/0	1	1	1	3	5	8	11	18	24	30	48	69
	2/0	1	1	1	3	4	6	9	14	19	25	40	57
	3/0	0	1	1	2	3	5	8	12	16	21	33	47
	4/0	0	1	1	1	2	4	6	10	13	17	27	39
Z	14	15	26	42	73	100	164	234	361	482	621	974	1405
	12	10	18	30	52	71	116	166	256	342	440	691	997
	10	6	11	18	32	43	71	102	157	209	269	423	610
	8	4	7	11	20	27	45	64	99	132	170	267	386
	6	3	5	8	14	19	31	45	69	93	120	188	271
	4	1	3	5	9	13	22	31	48	64	82	129	186
	3	1	2	4	7	9	16	22	35	47	60	94	136
	2	1	1	3	6	8	13	19	29	39	50	78	113
	1	1	1	2	5	6	10	15	23	31	40	63	92
XHH, XHHW, XHHW-2, ZW	14	9	15	25	44	59	98	140	216	288	370	581	839
	12	7	12	19	33	45	75	107	165	221	284	446	644
	10	5	9	14	25	34	56	80	123	164	212	332	480
	8	3	5	8	14	19	31	44	68	91	118	185	267
	6	1	3	6	10	14	23	33	51	68	87	137	197
	4	1	2	4	7	10	16	24	37	49	63	99	143
	3	1	1	3	6	8	14	20	31	41	53	84	121
	2	1	1	3	5	7	12	17	26	35	45	70	101
XHH, XHHW, XHHW-2	1	1	1	1	4	5	9	12	19	26	33	52	76
	1/0	1	1	1	3	4	7	10	16	22	28	44	64
	2/0	0	1	1	2	3	6	9	13	18	23	37	53
	3/0	0	1	1	1	3	5	7	11	15	19	30	44
	4/0	0	1	1	1	2	4	6	9	12	16	25	36
	250	0	0	1	1	1	3	5	7	10	13	20	30
	300	0	0	1	1	1	3	4	6	9	11	18	25
	350	0	0	1	1	1	2	3	6	7	10	15	22
	400	0	0	1	1	1	2	3	5	7	9	14	20
	500	0	0	0	1	1	1	2	4	5	7	11	16
	600	0	0	0	1	1	1	1	3	4	6	9	13
	700	0	0	0	1	1	1	1	3	4	5	8	11
	750	0	0	0	0	1	1	1	3	4	5	7	11

* Types RHH, RHW, AND RHW-2 without outer covering.

See Ugly's page 119 for Trade Size / Metric Designator conversion.

MAXIMUM NUMBER OF CONDUCTORS IN *FLEXIBLE METAL CONDUIT*

Type Letters	Cond. Size AWG/kcmil	Trade Sizes In Inches $^1/_2$	$^3/_4$	1	$1^1/_4$	$1^1/_2$	2	$2^1/_2$	3	$3^1/_2$	4
RHH, RHW, RHW-2	14	4	7	11	17	25	44	67	96	131	171
	12	3	6	9	14	21	37	55	80	109	142
	10	3	5	7	11	17	30	45	64	88	115
	8	1	2	4	6	9	15	23	34	46	60
	6	1	1	3	5	7	12	19	27	37	48
	4	1	1	2	4	5	10	14	21	29	37
	3	1	1	1	3	5	8	13	18	25	33
	2	1	1	1	3	4	7	11	16	22	28
	1	0	1	1	1	2	5	7	10	14	19
	1/0	0	1	1	1	2	4	6	9	12	16
	2/0	0	1	1	1	1	3	5	8	11	14
	3/0	0	0	1	1	1	3	5	7	9	12
	4/0	0	0	1	1	1	2	4	6	8	10
	250	0	0	0	1	1	1	3	4	6	8
	300	0	0	0	1	1	1	2	4	5	7
	350	0	0	0	1	1	1	2	3	5	6
	400	0	0	0	0	1	1	1	3	4	6
	500	0	0	0	0	1	1	1	3	4	5
	600	0	0	0	0	1	1	1	2	3	4
	700	0	0	0	0	0	1	1	1	3	3
	750	0	0	0	0	0	1	1	1	2	3
TW THHW, THW, THW-2	14	9	15	23	36	53	94	141	203	277	361
	12	7	11	18	28	41	72	108	156	212	277
	10	5	8	13	21	30	54	81	116	158	207
	8	3	5	7	11	17	30	45	64	88	115
RHH*, RHW*, RHW-2*	14	6	10	15	24	35	62	94	135	184	240
	12	5	8	12	19	28	50	75	108	148	193
	10	4	6	10	15	22	39	59	85	115	151
	8	1	4	6	9	13	23	35	51	69	90
RHH*, RHW*, RHW-2*, TW, THW, THHW, THW-2	6	1	3	4	7	10	18	27	39	53	69
	4	1	1	3	5	7	13	20	29	39	51
	3	1	1	3	4	6	11	17	25	34	44
	2	1	1	2	4	5	10	14	21	29	37
	1	1	1	1	2	4	7	10	15	20	26
	1/0	0	1	1	1	3	6	9	12	17	22
	2/0	0	1	1	1	3	5	7	10	14	19
	3/0	0	1	1	1	2	4	6	9	12	16
	4/0	0	0	1	1	1	3	5	7	10	13
	250	0	0	1	1	1	3	4	6	8	11
	300	0	0	1	1	1	2	3	5	7	9
	350	0	0	0	1	1	1	3	4	6	8
	400	0	0	0	1	1	1	3	4	6	7
	500	0	0	0	1	1	1	2	3	5	6
	600	0	0	0	0	1	1	1	3	4	5
	700	0	0	0	0	1	1	1	2	3	4
	750	0	0	0	0	1	1	1	2	3	4
THHN, THWN, THWN-2	14	13	22	33	52	76	134	202	291	396	518
	12	9	16	24	38	56	98	147	212	289	378
	10	6	10	15	24	35	62	93	134	182	238
	8	3	6	9	14	20	35	53	77	105	137
	6	2	4	6	10	14	25	38	55	76	99
	4	1	2	4	6	9	16	24	34	46	61
	3	1	1	3	5	7	13	20	29	39	51
	2	1	1	3	4	6	11	17	24	33	43
	1	1	1	1	3	4	8	12	18	24	32
	1/0	1	1	1	2	4	7	10	15	20	27
	2/0	0	1	1	1	3	6	9	12	17	22
	3/0	0	1	1	1	2	5	7	10	14	18

MAXIMUM NUMBER OF CONDUCTORS IN *FLEXIBLE METAL CONDUIT*

Type Letters	Cond. Size AWG/kcmil	Trade Sizes In Inches ½	¾	1	1¼	1½	2	2½	3	3½	4
THHN, THWN, THWN-2	4/0	0	1	1	1	1	4	6	8	12	15
	250	0	0	1	1	1	3	5	7	9	12
	300	0	0	1	1	1	3	4	6	8	11
	350	0	0	1	1	1	2	3	5	7	9
	400	0	0	0	1	1	1	3	5	6	8
	500	0	0	0	1	1	1	2	4	5	7
	600	0	0	0	0	1	1	1	3	4	5
	700	0	0	0	0	1	1	1	3	4	5
	750	0	0	0	0	1	1	1	2	3	4
FEP, FEPB, PFA, PFAH, TFE	14	12	21	32	51	74	130	196	282	385	502
	12	9	15	24	37	54	95	143	206	281	367
	10	6	11	17	26	39	68	103	148	201	263
	8	4	6	10	15	22	39	59	85	115	151
	6	2	4	7	11	16	28	42	60	82	107
	4	1	3	5	7	11	19	29	42	57	75
	3	1	2	4	6	9	16	24	35	48	62
	2	1	1	3	5	7	13	20	29	39	51
PFA, PFAH, TFE	1	1	1	2	3	5	9	14	20	27	36
PFA, PFAH, TFE, Z	1/0	1	1	1	3	4	8	11	17	23	30
	2/0	1	1	1	2	3	6	9	14	19	24
	3/0	0	1	1	1	3	5	8	11	15	20
	4/0	0	1	1	1	2	4	6	9	13	16
Z	14	15	25	39	61	89	157	236	340	463	605
	12	11	18	28	43	63	111	168	241	329	429
	10	6	11	17	26	39	68	103	148	201	263
	8	4	7	11	17	24	43	65	93	127	166
	6	3	5	7	12	17	30	45	65	89	117
	4	1	3	5	8	12	21	31	45	61	80
	3	1	2	4	6	8	15	23	33	45	58
	2	1	1	3	5	7	12	19	27	37	49
	1	1	1	2	4	6	10	15	22	30	39
XHH, XHHW, XHHW-2, ZW	14	9	15	23	36	53	94	141	203	277	361
	12	7	11	18	28	41	72	108	156	212	277
	10	5	8	13	21	30	54	81	116	158	207
	8	3	5	7	11	17	30	45	64	88	115
	6	1	3	5	8	12	22	33	48	65	85
	4	1	2	4	6	9	16	24	34	47	61
	3	1	1	3	5	7	13	20	29	40	52
	2	1	1	3	4	6	11	17	24	33	44
XHH, XHHW, XHHW-2	1	1	1	1	3	5	8	13	18	25	32
	1/0	1	1	1	2	4	7	10	15	21	27
	2/0	0	1	1	2	3	6	9	13	17	23
	3/0	0	1	1	1	3	5	7	10	14	19
	4/0	0	1	1	1	2	4	6	9	12	15
	250	0	0	1	1	1	3	5	7	10	13
	300	0	0	1	1	1	3	4	6	8	11
	350	0	0	1	1	1	2	4	5	7	9
	400	0	0	0	1	1	1	3	5	6	8
	500	0	0	0	1	1	1	3	4	5	7
	600	0	0	0	0	1	1	1	3	4	5
	700	0	0	0	0	1	1	1	3	4	5
	750	0	0	0	0	1	1	1	2	3	4

* Types RHH, RHW, AND RHW-2 without outer covering.

See Ugly's page 119 for Trade Size / Metric Designator conversion.

MAXIMUM NUMBER OF CONDUCTORS IN *LIQUIDTIGHT FLEXIBLE METAL CONDUIT*

Type Letters	Cond. Size AWG/kcmil	Trade Sizes In Inches $^{1}/_{2}$	$^{3}/_{4}$	1	$1^{1}/_{4}$	$1^{1}/_{2}$	2	$2^{1}/_{2}$	3	$3^{1}/_{2}$	4
RHH, RHW, RHW-2	14	4	7	12	21	27	44	66	102	133	173
	12	3	6	10	17	22	36	55	84	110	144
	10	3	5	8	14	18	29	44	68	89	116
	8	1	2	4	7	9	15	23	36	46	61
	6	1	1	3	6	7	12	18	28	37	48
	4	1	1	2	4	6	9	14	22	29	38
	3	1	1	1	4	5	8	13	19	25	33
	2	1	1	1	3	4	7	11	17	22	29
	1	0	1	1	1	3	5	7	11	14	19
	1/0	0	1	1	1	2	4	6	10	13	16
	2/0	0	1	1	1	1	3	5	8	11	14
	3/0	0	0	1	1	1	3	4	7	9	12
	4/0	0	0	1	1	1	2	4	6	8	10
	250	0	0	0	1	1	1	3	4	6	8
	300	0	0	0	1	1	1	2	4	5	7
	350	0	0	0	1	1	1	2	3	5	6
	400	0	0	0	1	1	1	1	3	4	6
	500	0	0	0	1	1	1	1	3	4	5
	600	0	0	0	0	1	1	1	2	3	4
	700	0	0	0	0	0	1	1	1	3	3
	750	0	0	0	0	0	1	1	1	2	3
TW THHW, THW, THW-2	14	9	15	25	44	57	93	140	215	280	365
	12	7	12	19	33	43	71	108	165	215	280
	10	5	9	14	25	32	53	80	123	160	209
	8	3	5	8	14	18	29	44	68	89	116
RHH*, RHW*, RHW-2*	14	6	10	16	29	38	62	93	143	186	243
	12	5	8	13	23	30	50	75	115	149	195
	10	3	6	10	18	23	39	58	89	117	152
	8	1	4	6	11	14	23	35	53	70	91
RHH*, RHW*, RHW-2*, TW, THW, THHW, THW-2	6	1	3	5	8	11	18	27	41	53	70
	4	1	1	3	6	8	13	20	30	40	52
	3	1	1	3	5	7	11	17	26	34	44
	2	1	1	2	4	6	9	14	22	29	38
	1	1	1	1	3	4	7	10	15	20	26
	1/0	0	1	1	2	3	6	8	13	17	23
	2/0	0	1	1	2	3	5	7	11	15	19
	3/0	0	1	1	1	2	4	6	9	12	16
	4/0	0	0	1	1	1	3	5	8	10	13
	250	0	0	1	1	1	3	4	6	8	11
	300	0	0	1	1	1	2	3	5	7	9
	350	0	0	0	1	1	1	3	5	6	8
	400	0	0	0	1	1	1	3	4	6	7
	500	0	0	0	1	1	1	2	3	5	6
	600	0	0	0	1	1	1	1	3	4	5
	700	0	0	0	0	1	1	1	2	3	4
	750	0	0	0	0	1	1	1	2	3	4
THHN, THWN, THWN-2	14	13	22	36	63	81	133	201	308	401	523
	12	9	16	26	46	59	97	146	225	292	381
	10	6	10	16	29	37	61	92	141	184	240
	8	3	6	9	16	21	35	53	81	106	138
	6	2	4	7	12	15	25	38	59	76	100
	4	1	2	4	7	9	15	23	36	47	61
	3	1	1	3	6	8	13	20	30	40	52
	2	1	1	3	5	7	11	17	26	33	44
	1	1	1	1	4	5	8	12	19	25	32
	1/0	1	1	1	3	4	7	10	16	21	27
	2/0	0	1	1	2	3	6	8	13	17	23
	3/0	0	1	1	1	3	5	7	11	14	19

MAXIMUM NUMBER OF CONDUCTORS IN *LIQUIDTIGHT FLEXIBLE METAL CONDUIT*

Type Letters	Cond. Size AWG/kcmil	Trade Sizes In Inches 1/2	3/4	1	1 1/4	1 1/2	2	2 1/2	3	3 1/2	4
THHN, THWN, THWN-2	4/0	0	1	1	1	2	4	6	9	12	15
	250	0	0	1	1	1	3	5	7	10	12
	300	0	0	1	1	1	3	4	6	8	11
	350	0	0	1	1	1	2	3	5	7	9
	400	0	0	0	1	1	1	3	5	6	8
	500	0	0	0	1	1	1	2	4	5	7
	600	0	0	0	1	1	1	1	3	4	6
	700	0	0	0	1	1	1	1	3	4	5
	750	0	0	0	0	1	1	1	3	3	5
FEP, FEPB, PFA, PFAH, TFE	14	12	21	35	61	79	129	195	299	389	507
	12	9	15	25	44	57	94	142	218	284	370
	10	6	11	18	32	41	68	102	156	203	266
	8	3	6	10	18	23	39	58	89	117	152
	6	2	4	7	13	17	27	41	64	83	108
	4	1	3	5	9	12	19	29	44	58	75
	3	1	2	4	7	10	16	24	37	48	63
	2	1	1	3	6	8	13	20	30	40	52
PFA, PFAH, TFE	1	1	1	2	4	5	9	14	21	28	36
PFA, PFAH, TFE, Z	1/0	1	1	1	3	4	7	11	18	23	30
	2/0	1	1	1	3	4	6	9	14	19	25
	3/0	0	1	1	2	3	5	8	12	16	20
	4/0	0	1	1	1	2	4	6	10	13	17
Z	14	20	26	42	73	95	156	235	360	469	611
	12	14	18	30	52	67	111	167	255	332	434
	10	8	11	18	32	41	68	102	156	203	266
	8	5	7	11	20	26	43	64	99	129	168
	6	4	5	8	14	18	30	45	69	90	118
	4	2	3	5	9	12	20	31	48	62	81
	3	2	2	4	7	9	15	23	35	45	59
	2	1	1	3	6	7	12	19	29	38	49
	1	1	1	2	5	6	10	15	23	30	40
XHH, XHHW, XHHW-2, ZW	14	9	15	25	44	57	93	140	215	280	365
	12	7	12	19	33	43	71	108	165	215	280
	10	5	9	14	25	32	53	80	123	160	209
	8	3	5	8	14	18	29	44	68	89	116
	6	1	3	6	10	13	22	33	50	66	86
	4	1	2	4	7	9	16	24	36	48	62
	3	1	1	3	6	8	13	20	31	40	52
	2	1	1	3	5	7	11	17	26	34	44
XHH, XHHW, XHHW-2	1	1	1	1	4	5	8	12	19	25	33
	1/0	1	1	1	3	4	7	10	16	21	28
	2/0	0	1	1	2	3	6	9	13	17	23
	3/0	0	1	1	1	3	5	7	11	14	19
	4/0	0	1	1	1	2	4	6	9	12	16
	250	0	0	1	1	1	3	5	7	10	13
	300	0	0	1	1	1	3	4	6	8	11
	350	0	0	1	1	1	2	3	5	7	10
	400	0	0	0	1	1	1	3	5	6	8
	500	0	0	0	1	1	1	2	4	5	7
	600	0	0	0	1	1	1	1	3	4	6
	700	0	0	0	1	1	1	1	3	4	5
	750	0	0	0	0	1	1	1	3	3	5

* Types RHH, RHW, AND RHW-2 without outer covering.

See Ugly's page 119 for Trade Size / Metric Designator conversion.

DIMENSIONS OF INSULATED CONDUCTORS & FIXTURE WIRES

TYPE	SIZE	APPROX. AREA SQ. IN.
RFH-2	18	0.0145
FFH-2	16	0.0172
RHW-2, RHH	14	0.0293
RHW	12	0.0353
	10	0.0437
	8	0.0835
	6	0.1041
	4	0.1333
	3	0.1521
	2	0.1750
	1	0.2660
	1/0	0.3039
	2/0	0.3505
	3/0	0.4072
	4/0	0.4754
	250	0.6291
	300	0.7088
	350	0.7870
	400	0.8626
	500	1.0082
	600	1.2135
	700	1.3561
	750	1.4272
	800	1.4957
	900	1.6377
	1000	1.7719
	1250	2.3479
	1500	2.6938
	1750	3.0357
	2000	3.3719
SF-2, SFF-2	18	0.0115
	16	0.0139
	14	0.0172
SF-1, SFF-1	18	0.0065
RFH-1, XF, XFF	18	0.0080
TF, TFF, XF, XFF	16	0.0109
TW, XF, XFF, THHW, THW, THW-2	14	0.0139
TW, THHW,	12	0.0181
THW, THW-2	10	0.0243
	8	0.0437
RHH*, RHW*, RHW-2*,	14	0.0209
	12	0.0260

*Types RHH, RHW, and RHW-2 without outer covering

TYPE	SIZE	APPROX. AREA SQ. IN.
THHW, THW, AF, XF, XFF	10	0.0333
RHH*, RHW*, RHW-2*	8	0.0556
TW, THW	6	0.0726
THHW	4	0.0973
THW-2	3	0.1134
RHH*	2	0.1333
RHW*	1	0.1901
RHW-2*	1/0	0.2223
	2/0	0.2624
	3/0	0.3117
	4/0	0.3718
	250	0.4596
	300	0.5281
	350	0.5958
	400	0.6619
	500	0.7901
	600	0.9729
	700	1.1010
	750	1.1652
	800	1.2272
	900	1.3561
	1000	1.4784
	1250	1.8602
	1500	2.1695
	1750	2.4773
	2000	2.7818
TFN	18	0.0055
TFFN	16	0.0072
THHN	14	0.0097
THWN	12	0.0133
THWN-2	10	0.0211
	8	0.0366
	6	0.0507
	4	0.0824
	3	0.0973
	2	0.1158
	1	0.1562
	1/0	0.1855
	2/0	0.2223
	3/0	0.2679
	4/0	0.3237
	250	0.3970
	300	0.4608
	350	0.5242
	400	0.5863
	500	0.7073
	600	0.8676
	700	0.9887

DIMENSIONS OF INSULATED CONDUCTORS & FIXTURE WIRES

TYPE	SIZE	APPROX. AREA SQ. IN.
THHN THWN THWN-2	750	1.0496
	800	1.1085
	900	1.2311
	1000	1.3478
PF, PGFF, PGF, PFF, PTF, PAF, PTFF, PAFF	18	0.0058
	16	0.0075
PF, PGFF, PGF, PFF, PTF, PAF, PTFF, PAFF TFE, FEP, PFA FEPB, PFAH	14	0.0100
TFE, FEP, PFA, FEPB, PFAH	12	0.0137
	10	0.0191
	8	0.0333
	6	0.0468
	4	0.0670
	3	0.0804
	2	0.0973
TFE, PFAH	1	0.1399
TFE, PFA, PFAH, Z	1/0	0.1676
	2/0	0.2027
	3/0	0.2463
	4/0	0.3000
ZF, ZFF	18	0.0045
	16	0.0061
Z, ZF, ZFF	14	0.0083
Z	12	0.0117
	10	0.0191
	8	0.0302
	6	0.0430
	4	0.0625
	3	0.0855
	2	0.1029
	1	0.1269
XHHW, ZW XHHW-2 XHH	14	0.0139
	12	0.0181
	10	0.0243
	8	0.0437
	6	0.0590
	4	0.0814
	3	0.0962
	2	0.1146
XHHW XHHW-2 XHH	1	0.1534
	1/0	0.1825
	2/0	0.2190
	3/0	0.2642
	4/0	0.3197
	250	0.3904

TYPE	SIZE	APPROX. AREA SQ. IN.
XHHW XHHW-2 XHH	300	0.4536
	350	0.5166
	400	0.5782
	500	0.6984
	600	0.8709
	700	0.9923
	750	1.0532
	800	1.1122
	900	1.2351
	1000	1.3519
	1250	1.7180
	1500	2.0157
	1750	2.3127
	2000	2.6073
KF-2 KFF-2	18	0.0031
	16	0.0044
	14	0.0064
	12	0.0093
	10	0.0139
KF-1 KFF-1	18	0.0026
	16	0.0037
	14	0.0055
	12	0.0083
	10	0.0127

*Types RHH, RHW, and RHW-2 without outer covering

See Ugly's page 117 - 123 for conversion of Square inches to mm^2.

COMPACT ALUMINUM BUILDING WIRE NOMINAL DIMENSIONS* AND AREAS

Bare Conductor			Types THW and THHW		Type THHN		Type XHHW		
Size AWG or kcmil	Number of Strands	Diam. Inches	Approx. Diam. Inches	Approx. Area Sq. In.	Approx. Diam. Inches	Approx. Area Sq. In.	Approx. Diam. Inches	Approx. Area Sq. Inches	Size AWG or kcmil
8	7	0.134	0.255	0.0510	—	—	0.224	0.0394	8
6	7	0.169	0.290	0.0660	0.240	0.0452	0.260	0.0530	6
4	7	0.213	0.335	0.0881	0.305	0.0730	0.305	0.0730	4
2	7	0.268	0.390	0.1194	0.360	0.1017	0.360	0.1017	2
1	19	0.299	0.465	0.1698	0.415	0.1352	0.415	0.1352	1
1/0	19	0.336	0.500	0.1963	0.450	0.1590	0.450	0.1590	1/0
2/0	19	0.376	0.545	0.2332	0.495	0.1924	0.490	0.1885	2/0
3/0	19	0.423	0.590	0.2733	0.540	0.2290	0.540	0.2290	3/0
4/0	19	0.475	0.645	0.3267	0.595	0.2780	0.590	0.2733	4/0
250	37	0.520	0.725	0.4128	0.670	0.3525	0.660	0.3421	250
300	37	0.570	0.775	0.4717	0.720	0.4071	0.715	0.4015	300
350	37	0.616	0.820	0.5281	0.770	0.4656	0.760	0.4536	350
400	37	0.659	0.865	0.5876	0.815	0.5216	0.800	0.5026	400
500	37	0.736	0.940	0.6939	0.885	0.6151	0.880	0.6082	500
600	61	0.813	1.050	0.8659	0.985	0.7620	0.980	0.7542	600
700	61	0.877	1.110	0.9676	1.050	0.8659	1.050	0.8659	700
750	61	0.908	1.150	1.0386	1.075	0.9076	1.090	0.9331	750
1000	61	1.060	1.285	1.2968	1.255	1.2370	1.230	1.1882	1000

* Dimensions are from industry sources.

See Ugly's page 117 - 122 for metric conversions.

DIMENSIONS AND PERCENT AREA OF CONDUIT AND TUBING

(for the combinations of wires permitted in Table 1, Chapter 9, *NEC* ©)

(See Ugly's page 117 - 123 for metric conversion)

Trade Size Inches	Internal Diameter Inches	Total Area 100% Sq. in.	2 Wires 31% Sq. in.	Over 2 Wires 40% Sq. in.	1 Wire 53% Sq. in.	60% Sq. in.
ELECTRICAL METALLIC TUBING (EMT)						
1/2	0.622	0.304	0.094	0.122	0.161	0.182
3/4	0.824	0.533	0.165	0.213	0.283	0.320
1	1.049	0.864	0.268	0.346	0.458	0.519
1 1/4	1.380	1.496	0.464	0.598	0.793	0.897
1 1/2	1.610	2.036	0.631	0.814	1.079	1.221
2	2.067	3.356	1.040	1.342	1.778	2.013
2 1/2	2.731	5.858	1.816	2.343	3.105	3.515
3	3.356	8.846	2.742	3.538	4.688	5.307
3 1/2	3.834	11.545	3.579	4.618	6.119	6.927
4	4.334	14.753	4.573	5.901	7.819	8.852
ELECTRICAL NONMETALLIC TUBING (ENT)						
1/2	0.560	0.246	0.076	0.099	0.131	0.148
3/4	0.760	0.454	0.141	0.181	0.240	0.272
1	1.000	0.785	0.243	0.314	0.416	0.471
1 1/4	1.340	1.410	0.437	0.564	0.747	0.846
1 1/2	1.570	1.936	0.600	0.774	1.026	1.162
2	2.020	3.205	0.993	1.282	1.699	1.923
2 1/2	–	–	–	–	–	–
3	–	–	–	–	–	–
3 1/2	–	–	–	–	–	–
4	–	–	–	–	–	–
FLEXIBLE METAL CONDUIT (FMT)						
3/8	0.384	0.116	0.036	0.046	0.061	0.069
1/2	0.635	0.317	0.098	0.127	0.168	0.190
3/4	0.824	0.533	0.165	0.213	0.283	0.320
1	1.020	0.817	0.253	0.327	0.433	0.490
1 1/4	1.275	1.277	0.396	0.511	0.677	0.766
1 1/2	1.538	1.858	0.576	0.743	0.985	1.115
2	2.040	3.269	1.013	1.307	1.732	1.961
2 1/2	2.500	4.909	1.522	1.963	2.602	2.945
3	3.000	7.069	2.191	2.827	3.746	4.241
3 1/2	3.500	9.621	2.983	3.848	5.099	5.773
4	4.000	12.566	3.896	5.027	6.660	7.540
INTERMEDIATE METAL CONDUIT (IMC)						
3/8	–	–	–	–	–	–
1/2	0.660	0.342	0.106	0.137	0.181	0.205
3/4	0.864	0.586	0.182	0.235	0.311	0.352
1	1.105	0.959	0.297	0.384	0.508	0.575
1 1/4	1.448	1.647	0.510	0.659	0.873	0.988
1 1/2	1.683	2.225	0.690	0.890	1.179	1.335
2	2.150	3.630	1.125	1.452	1.924	2.178
2 1/2	2.557	5.135	1.592	2.054	2.722	3.081
3	3.176	7.922	2.456	3.169	4.199	4.753
3 1/2	3.671	10.584	3.281	4.234	5.610	6.351
4	4.166	13.631	4.226	5.452	7.224	8.179

DIMENSIONS AND PERCENT AREA OF CONDUIT AND TUBING

(for the combinations of wires permitted in Table 1, Chapter 9, *NEC*©)

(See Ugly's page 117 - 123 for metric conversion)

Trade Size Inches	Internal Diameter Inches	Total Area 100% Sq. in.	2 Wires 31% Sq. in.	Over 2 Wires 40% Sq. in.	1 Wire 53% Sq. in.	60% Sq. in.
LIQUIDTIGHT FLEXIBLE NONMETALLIC CONDUIT (TYPE LFNC-B*)						
3/8	0.494	0.192	0.059	0.077	0.102	0.115
1/2	0.632	0.314	0.097	0.125	0.166	0.188
3/4	0.830	0.541	0.168	0.216	0.287	0.325
1	1.054	0.873	0.270	0.349	0.462	0.524
1 1/4	1.395	1.528	0.474	0.611	0.810	0.917
1 1/2	1.588	1.981	0.614	0.792	1.050	1.188
2	2.033	3.246	1.006	1.298	1.720	1.948

* Corresponds to Section 356.2(2).

Trade Size Inches	Internal Diameter Inches	Total Area 100% Sq. in.	2 Wires 31% Sq. in.	Over 2 Wires 40% Sq. in.	1 Wire 53% Sq. in.	60% Sq. in.
LIQUIDTIGHT FLEXIBLE NONMETALLIC CONDUIT (TYPE LFNC-A*)						
3/8	0.495	0.192	0.060	0.077	0.102	0.115
1/2	0.630	0.312	0.097	0.125	0.165	0.187
3/4	0.825	0.535	0.166	0.214	0.283	0.321
1	1.043	0.854	0.265	0.342	0.453	0.513
1 1/4	1.383	1.502	0.466	0.601	0.796	0.901
1 1/2	1.603	2.018	0.626	0.807	1.070	1.211
2	2.063	3.343	1.036	1.337	1.772	2.006

* Corresponds to Section 356.2(1).

Trade Size Inches	Internal Diameter Inches	Total Area 100% Sq. in.	2 Wires 31% Sq. in.	Over 2 Wires 40% Sq. in.	1 Wire 53% Sq. in.	60% Sq. in.
LIQUIDTIGHT FLEXIBLE METAL CONDUIT (LFMC)						
3/8	0.494	0.192	0.059	0.077	0.102	0.115
1/2	0.632	0.314	0.097	0.125	0.166	0.188
3/4	0.830	0.541	0.168	0.216	0.287	0.325
1	1.054	0.873	0.270	0.349	0.462	0.524
1 1/4	1.395	1.528	0.474	0.611	0.810	0.917
1 1/2	1.588	1.981	0.614	0.792	1.050	1.188
2	2.033	3.246	1.006	1.298	1.720	1.948
2 1/2	2.493	4.881	1.513	1.953	2.587	2.929
3	3.085	7.475	2.317	2.990	3.962	4.485
3 1/2	3.520	9.731	3.017	3.893	5.158	5.839
4	4.020	12.692	3.935	5.077	6.727	7.615
RIGID METAL CONDUIT (RMC)						
3/8	–	–	–	–	–	–
1/2	0.632	0.314	0.097	0.125	0.166	0.188
3/4	0.836	0.549	0.170	0.220	0.291	0.329
1	1.063	0.887	0.275	0.355	0.470	0.532
1 1/4	1.394	1.526	0.473	0.610	0.809	0.916
1 1/2	1.624	2.071	0.642	0.829	1.098	1.243
2	2.083	3.408	1.056	1.363	1.806	2.045
2 1/2	2.489	4.866	1.508	1.946	2.579	2.919
3	3.090	7.499	2.325	3.000	3.974	4.499
3 1/2	3.570	10.010	3.103	4.004	5.305	6.006
4	4.050	12.882	3.994	5.153	6.828	7.729
5	5.073	20.212	6.266	8.085	10.713	12.127
6	6.093	29.158	9.039	11.663	15.454	17.495

DIMENSIONS AND PERCENT AREA OF CONDUIT AND TUBING

(for the combinations of wires permitted in Table 1, Chapter 9, *NEC*®)

(See Ugly's page 117 - 123 for metric conversion)

RIGID PVC CONDUIT (RNC), SCHEDULE 80

Trade Size Inches	Internal Diameter Inches	Total Area 100% Sq. in.	2 Wires 31% Sq. in.	Over 2 Wires 40% Sq. in.	1 Wire 53% Sq. in.	60% Sq. in.
1/2	0.526	0.217	0.067	0.087	0.115	0.130
3/4	0.722	0.409	0.127	0.164	0.217	0.246
1	0.936	0.688	0.213	0.275	0.365	0.413
1 1/4	1.255	1.237	0.383	0.495	0.656	0.742
1 1/2	1.476	1.711	0.530	0.684	0.907	1.027
2	1.913	2.874	0.891	1.150	1.523	1.725
2 1/2	2.290	4.119	1.277	1.647	2.183	2.471
3	2.864	6.442	1.997	2.577	3.414	3.865
3 1/2	3.326	8.688	2.693	3.475	4.605	5.213
4	3.786	11.258	3.490	4.503	5.967	6.755
5	4.768	17.855	5.535	7.142	9.463	10.713
6	5.709	25.598	7.935	10.239	13.567	15.359

RIGID PVC CONDUIT (RNC), SCHEDULE 40 & HDPE CONDUIT

Trade Size Inches	Internal Diameter Inches	Total Area 100% Sq. in.	2 Wires 31% Sq. in.	Over 2 Wires 40% Sq. in.	1 Wire 53% Sq. in.	60% Sq. in.
1/2	0.602	0.285	0.088	0.114	0.151	0.171
3/4	0.804	0.508	0.157	0.203	0.269	0.305
1	1.029	0.832	0.258	0.333	0.441	0.499
1 1/4	1.360	1.453	0.450	0.581	0.770	0.872
1 1/2	1.590	1.986	0.616	0.794	1.052	1.191
2	2.047	3.291	1.020	1.316	1.744	1.975
2 1/2	2.445	4.695	1.455	1.878	2.488	2.817
3	3.042	7.268	2.253	2.907	3.852	4.361
3 1/2	3.521	9.737	3.018	3.895	5.161	5.842
4	3.998	12.554	3.892	5.022	6.654	7.532
5	5.016	19.761	6.126	7.904	10.473	11.856
6	6.031	28.567	8.856	11.427	15.141	17.140

TYPE A, RIGID PVC CONDUIT (RNC)

Trade Size Inches	Internal Diameter Inches	Total Area 100% Sq. in.	2 Wires 31% Sq. in.	Over 2 Wires 40% Sq. in.	1 Wire 53% Sq. in.	60% Sq. in.
1/2	0.700	0.385	0.119	0.154	0.204	0.231
3/4	0.910	0.650	0.202	0.260	0.345	0.390
1	1.175	1.084	0.336	0.434	0.575	0.651
1 1/4	1.500	1.767	0.548	0.707	0.937	1.060
1 1/2	1.720	2.324	0.720	0.929	1.231	1.394
2	2.155	3.647	1.131	1.459	1.933	2.188
2 1/2	2.635	5.453	1.690	2.181	2.890	3.272
3	3.230	8.194	2.540	3.278	4.343	4.916
3 1/2	3.690	10.694	3.315	4.278	5.668	6.416
4	4.180	13.723	4.254	5.489	7.273	8.234

TYPE EB, PVC CONDUIT (RNC)

Trade Size Inches	Internal Diameter Inches	Total Area 100% Sq. in.	2 Wires 31% Sq. in.	Over 2 Wires 40% Sq. in.	1 Wire 53% Sq. in.	60% Sq. in.
2	2.221	3.874	1.201	1.550	2.053	2.325
2 1/2	–	–		–	–	–
3	3.330	8.709	2.700	3.484	4.616	5.226
3 1/2	3.804	11.365	3.523	4.546	6.023	6.819
4	4.289	14.448	4.479	5.779	7.657	8.669
5	5.316	22.195	6.881	8.878	11.763	13.317
6	6.336	31.530	9.774	12.612	16.711	18.918

THREAD DIMENSIONS AND TAP DRILL SIZES

COARSE THREAD SERIES				FINE THREAD SERIES			
NOMINAL SIZE	THREADS PER IN.	TAP DRILL	CLEARANCE DRILL	NOMINAL SIZE	THREADS PER IN.	TAP DRILL	CLEARANCE DRILL
5/64"	48	47	36	0	80	3/64"	51
1/8	40	38	29	1	72	53	47
6	32	36	25	2	64	50	42
8	32	29	16	3	56	45	36
10	24	25	13/64"	4	48	42	31
12	24	16	7/32"	1/8"	44	37	29
1/4"	20	7	17/64"	6	40	33	25
5/16"	18	F	21/64"	8	36	29	16
3/8"	16	5/16"	25/64"	10	32	21	13/64"
7/16"	14	U	29/64"	12	28	14	7/32"
1/2"	13	27/64"	33/64"	1/4"	28	3	17/64"
9/16"	12	31/64"	37/64"	5/16"	24	1	21/64"
5/8"	11	17/32"	41/64"	3/8"	24	Q	25/64"
3/4"	10	21/32"	49/64"	7/16"	20	25/64"	29/64"
7/8"	9	49/64"	57/64"	1/2"	20	29/64"	33/64"
1"	8	7/8"	1-1/64"	9/16"	18	33/64"	37/64"
1-1/4"	7	1-11/64"	1-17/64"	5/8"	18	37/64"	41/64"
1-3/8"	6	1-19/64"	1-25/64"	3/4"	16	11/16"	49/64"
1-1/2"	6	1-27/64"	1-33/64"	7/8"	14	13/16"	57/64"
2"	4-1/2"	1-25/32"	2-1/32"	1	14	15/16"	1-1/64"

HOLE SAW CHART

TRADE SIZE	RIGID CONDUIT	E.M.T. CONDUIT	GREEN-FIELD	L.T. FLEX.	TRADE SIZE	RIGID CONDUIT	E.M.T. CONDUIT	GREEN-FIELD
1/2"	7/8"	3/4"	1"	1-1/8"	2-1/2"	3"	2-7/8"	2-7/8"
3/4	1-1/8"	1"	1-1/8"	1-1/4"	3"	3-5/8"	3-1/2"	3-5/8"
1"	1-3/8"	1-1/4"	1-1/2"	1-1/2"	3-1/2"	4-1/8"	4"	4-1/8"
1-1/4"	1-3/4"	1-5/8"	1-3/4"	1-7/8"	4"	4-5/8"	4-1/2"	4-5/8"
1-1/2"	2"	1-7/8"	2"	2-1/8"	5"	5-3/4"		
2"	2-1/2"	2-1/8"	2-1/2"	2-3/4"	6"	6-3/4"		

NOTE: For oil type push button station, use size 1-7/32" knock-out punch.

METAL BOXES

BOX DIMENSION, INCHES TRADE SIZE OR TYPE	MIN. CU. IN. CAPACITY	MAXIMUM NUMBER OF CONDUCTORS						
		NO. 18	NO. 16	NO. 14	NO. 12	NO. 10	NO. 8	NO. 6
4 x 1-1/4 ROUND OR OCTAGONAL	12.5	8	7	6	5	5	4	2
4 x 1-1/2 ROUND OR OCTAGONAL	15.5	10	8	7	6	6	5	3
4 x 2-1/8 ROUND OR OCTAGONAL	21.5	14	12	10	9	8	7	4
4 x 1-1/4 SQUARE	18.0	12	10	9	8	7	6	3
4 x 1-1/2 SQUARE	21.0	14	12	10	9	8	7	4
4 x 2-1/8 SQUARE	30.3	20	17	15	13	12	10	6
4-11/16 x 1-1/4 SQUARE	25.5	17	14	12	11	10	8	5
4-11/16 x 1-1/2 SQUARE	29.5	19	16	14	13	11	9	5
4-11/16 x 2-1/8 SQUARE	42.0	28	24	21	18	16	14	8
3 x 2 x 1-1/2 DEVICE	7.5	5	4	3	3	3	2	1
3 x 2 x 2 DEVICE	10.0	6	5	5	4	4	3	2
3 x 2 x 2-1/4 DEVICE	10.5	7	6	5	4	4	3	2
3 x 2 x 2-1/2 DEVICE	12.5	8	7	6	5	5	4	2
3 x 2 x 2-3/4 DEVICE	14.0	9	8	7	6	5	4	2
3 x 2 x 3-1/2 DEVICE	18.0	12	10	9	8	7	6	3
4 x 2-1/8 x 1-1/2 DEVICE	10.3	6	5	5	4	4	3	2
4 x 2-1/8 x 1-7/8 DEVICE	13.0	8	7	6	5	5	4	2
4 x 2-1/8 x 2-1/8 DEVICE	14.5	9	8	7	6	5	4	2
3-3/4 x 2 x 2-1/2 MASONRY BOX/GANG	14.0	9	8	7	6	5	4	2
3-3/4 x 2 x 3-1/2 MASONRY BOX/GANG	21.0	14	12	10	9	8	7	4
FS-MINIMUM INTERNAL DEPTH 1-3/4 SINGLE COVER/GANG	13.5	9	7	6	6	5	4	2
FD-MINIMUM INTERNAL DEPTH 2-3/8 SINGLE COVER/GANG	18.0	12	10	9	8	7	6	3
FS-MINIMUM INTERNAL DEPTH 1-3/4 MULTIPLE COVER/GANG	18.0	12	10	9	8	7	6	3
FD-MINIMUM INTERNAL DEPTH 2-3/8 MULTIPLE COVER/GANG	24.0	16	13	12	10	9	8	4

MINIMUM COVER REQUIREMENTS 0 - 600 VOLTS, NOMINAL

Cover is defined as the distance between the top surface of direct burial cable, conduit, or other raceways and the finished surface.

WIRING METHOD	MINIMUM BURIAL (INCHES)
DIRECT BURIAL CABLES	24
RIGID METAL CONDUIT	6
INTERMEDIATE METAL CONDUIT	6
RIGID NONMETALLIC CONDUIT (APPROVED FOR DIRECT BURIAL WITHOUT CONCRETE ENCASEMENT)	18

For most locations, for complete details, refer to National Electrical Code® Table 300-5 for exceptions such as highways, airports, driveways, parking lots, etc. See Ugly's page 117 - 123 for metric conversions.

VOLUME REQUIRED PER CONDUCTOR

SIZE OF CONDUCTOR	FREE SPACE WITHIN BOX FOR EACH CONDUCTOR
No. 18	1.5 CUBIC INCHES
No. 16	1.75 CUBIC INCHES
No. 14	2 CUBIC INCHES
No. 12	2.25 CUBIC INCHES
No. 10	2.5 CUBIC INCHES
No. 8	3 CUBIC INCHES
No. 6	5 CUBIC INCHES

For complete details see NEC 314.16B. See Ugly's page 117 - 123 for metric conversions.

VERTICAL CONDUCTOR SUPPORTS

AWG or Circular-Mil Size of Wire	Support of Conductors in Vertical Raceways	CONDUCTORS Aluminum or Copper-Clad Aluminum	Copper
18 AWG through 8 AWG	Not greater than	100 feet	100 feet
6 AWG through 1/0 AWG	Not greater than	200 feet	100 feet
2/0 AWG through 4/0 AWG	Not greater than	180 feet	80 feet
Over 4/0 AWG through 350 kcmil	Not greater than	135 feet	60 feet
Over 350 kcmil through 500 kcmil	Not greater than	120 feet	50 feet
Over 500 kcmil through 750 kcmil	Not greater than	95 feet	40 feet
Over 750 kcmil	Not greater than	85 feet	35 feet

For SI units: one foot = 0.3048 meter. See Ugly's page 117 - 123 for metric conversions.

MINIMUM DEPTH OF CLEAR WORKING SPACE IN FRONT OF ELECTRICAL EQUIPMENT

NOMINAL VOLTAGE TO GROUND	CONDITIONS		
	1	2	3
	Minimum clear distance (feet)		
0 - 150	3	3	3
151 - 600	3	3-1/2	4
601 - 2500	3	4	5
2501 - 9000	4	5	6
9001 - 25,000	5	6	9
25,001 - 75 kV	6	8	10
Above 75 kV	8	10	12

NOTES:

1. For SI units, 1 ft. = 0.3048 m.
2. Where the conditions are as follows:

* **Condition 1 –** Exposed live parts on one side and no live or grounded parts on the other side of the working space, or exposed live parts on both sides effectively guarded by suitable wood or other insulating materials. Insulated wire or insulated busbars operating at not over 300 volts shall not be considered live parts.

Condition 2 – Exposed live parts on one side and grounded parts on the other side. Concrete, brick, or tile walls shall be considered as grounded surfaces.

Condition 3 – Exposed live parts on both sides of the work space (not guarded as provided in Condition 1) with the operator between.

* 300 volts to ground for 600 volts and under.

See Ugly's page 117 - 123 for metric conversion.

MINIMUM CLEARANCE OF LIVE PARTS

NOMINAL VOLTAGE RATING KV	IMPULSE WITHSTAND B.I.L. KV		*MINIMUM CLEARANCE OF LIVE PARTS, INCHES			
			PHASE-TO-PHASE		PHASE-TO-GROUND	
	INDOORS	OUTDOORS	INDOORS	OUTDOORS	INDOORS	OUTDOORS
2.4 - 4.16	60	95	4.5	7	3.0	6
7.2	75	95	5.5	7	4.0	6
13.8	95	110	7.5	12	5.0	7
14.4	110	110	9.0	12	6.5	7
23	125	150	10.5	15	7.5	10
34.5	150	150	12.5	15	9.5	10
	200	200	18.0	18	13.0	13
46		200		18		13
		250		21		17
69		250		21		17
		350		31		25
115		550		53		42
138		550		53		42
		650		63		50
161		650		63		50
		750		72		58
230		750		72		58
		900		89		71
		1050		105		83

For SI units: one inch = 25.4 millimeters

* The values given are the minimum clearance for rigid parts and bare conductors under favorable service conditions. They shall be increased for conductor movement or under unfavorable service conditions, or wherever space limitations permit. The selection of the associated impulse withstand voltage for a particular system voltage is determined by the characteristics of the surge protective equipment.

See Ugly's page 117 - 123 for metric conversion.

MINIMUM SIZE EQUIPMENT GROUNDING CONDUCTORS FOR GROUNDING RACEWAY AND EQUIPMENT

RATING OR SETTING OF AUTOMATIC OVERCURRENT DEVICE IN CIRCUIT AHEAD OF EQUIPMENT, CONDUIT, ETC., NOT EXCEEDING (AMPERES)	SIZE	
	COPPER	ALUMINUM OR COPPER-CLAD ALUMINUM*
15	14	12
20	12	10
30	10	8
40	10	8
60	10	8
100	8	6
200	6	4
300	4	2
400	3	1
500	2	1/0
600	1	2/0
800	1/0	3/0
1000	2/0	4/0
1200	3/0	250 kcmil
1600	4/0	350 kcmil
2000	250 kcmil	400 kcmil
2500	350 kcmil	600 kcmil
3000	400 kcmil	600 kcmil
4000	500 kcmil	800 kcmil
5000	700 kcmil	1200 kcmil
6000	800 kcmil	1200 kcmil

NOTE: Where necessary to comply with Section 250.4(A)(5) or 250.4(B)(4), the equipment grounding conductor shall be sized larger than given in this table.
* See installation restrictions in NEC 250.120.

GROUNDING ELECTRODE CONDUCTOR FOR ALTERNATING-CURRENT SYSTEMS

SIZE OF LARGEST UNGROUNDED SERVICE-ENTRANCE CONDUCTOR OR EQUIVALENT AREA FOR PARALLEL CONDUCTORS*		SIZE OF GROUNDING ELECTRODE CONDUCTOR	
COPPER	ALUMINUM OR COPPER-CLAD ALUMINUM	COPPER	ALUMINUM OR COPPER-CLAD ALUMINUM**
2 OR SMALLER	1/0 OR SMALLER	8	6
1 OR 1/0	2/0 OR 3/0	6	4
2/0 OR 3/0	4/0 OR 250 kcmil	4	2
OVER 3/0 THRU 350kcmil	OVER 250 THRU THRU 500 kcmil	2	1/0
OVER 350 kcmil THRU 600 kcmil	OVER 500 kcmil THRU 900 kcmil	1/0	3/0
OVER 600 kcmil THRU 1100 kcmil	OVER 900 kcmil THRU 1750 kcmil	2/0	4/0
OVER 1100 kcmil	OVER 1750 kcmil	3/0	250 kcmil

NOTES:

1. Where multiple sets of service-entrance conductors are used as permit ted in Section 230.40, Exception No. 2, the equivalent size of the largest service-entrance conductor shall be determined by the largest sum of the areas of the corresponding conductors of each set.
2. Where there are no service-entrance conductors, the grounding electrode conductor size shall be determined by the equivalent size of the largest service-entrance conductor required for the load to be served.

*This table also applies to the derived conductors of separately derived ac systems.

**See installation restrictions in Section 250.64(a).

FPN: See Section 250-24(b) for size of ac system conductor brought to service equipment.

GENERAL LIGHTING LOADS BY OCCUPANCY

TYPE OF OCCUPANCY	VOLT-AMPERES PER SQUARE FOOT
Armories & auditoriums	1
Banks	3 1/2[b]
Barber shops & beauty parlors	3
Churches	1
Clubs	2
Court rooms	2
Dwelling units[a]	3
Garages - commercial (storage)	1/2
Hospitals	2
Hotels & motels, including apartment houses without provision for cooking by tenants[a]	2
Industrial commercial (loft) buildings	2
Lodge rooms	1½
Office buildings	3 1/2[b]
Restaurants	2
Schools	3
Stores	3
Warehouses (storage)	1/4
In any of the preceding occupancies except one-family dwellings & individual dwelling units of two-family & multi-family dwellings:	
Assembly halls & auditoriums	1
Halls, corridors, closets, stairways	1/2
Storage spaces	1/4

[a] See 220.3(B)(10)
[b] In addition, a unit load of 11 volt-amperes/m^2 or 1 volt-ampere/ft^2 shall be included for general-purpose receptacle outlets where the actual number of general-purpose receptacle outlets is unknown.

LIGHTING LOAD DEMAND FACTORS

Type of Occupancy	Portion of Lighting Load to Which Demand Factor Applies (Volt-Amperes)	Demand Factor (Percent)
Dwelling units	First 3000 or less at	100
	From 3001 to 120,000 at	35
	Remainder over 120,000 at	25
Hospitals*	First 50,000 or less at	40
	Remainder over 50,000 at	20
Hotels and motels, including apartment houses without provision for cooking by tenants*	First 20,000 or less at	50
	From 20,001 to 100,000 at	40
	Remainder over 100,000 at	30
Warehouses (Storage)	First 12,500 or less at	100
	Remainder over 12,500 at	50
All others	Total volt-amperes	100

* The demand factors of this table shall not apply to the computed load of feeders or services supplying areas in hospitals, hotels, and motels where the entire lighting is likely to be used at one time, as in operating rooms, ballrooms, or dining rooms.

DEMAND FACTORS FOR NONDWELLING RECEPTACLE LOADS

Portion of Receptacle Load to Which Demand Factor Applies (Volt-Amperes)	Demand Factor (Percent)
First 10 kVA or less at	100
Remainder over 10 kVA at	50

DEMAND FACTORS FOR HOUSEHOLD ELECTRIC CLOTHES DRYERS

Number of Dryers	Demand Factor (Percent)
1 - 4	100%
5	85%
6	75%
7	65%
8	60%
9	55%
10	50%
11	47%
12 - 22	% = 47 - (number of dryers - 11)
23	35%
24 - 42	% = 35 - [0.5 x (number of dryers - 23)]
43 and over	25%

DEMAND FACTORS FOR KITCHEN EQUIPMENT - OTHER THAN DWELLING UNIT(S)

Number of Units of Equipment	Demand Factor (Percent)
1	100
2	100
3	90
4	80
5	70
6 and over	65

DEMAND LOADS FOR HOUSEHOLD ELECTRIC RANGES, WALL-MOUNTED OVENS, COUNTER-MOUNTED COOKING UNITS, and OTHER HOUSEHOLD COOKING APPLIANCES over 1 3/4 kW RATING

(Column C to be used in all cases except as otherwise permitted in note 3)

Number of Appliances	Demand Factor (Percent) (See Notes)		Column C Maximum Demand (kW) (See Notes) (not over 12 kW Rating)
	Column A (less than 3 1/2 kW Rating)	Column B (3 1/2 kW to 8 3/4 kW Rating)	
1	80	80	8
2	75	65	11
3	70	55	14
4	66	50	17
5	62	45	20
6	59	43	21
7	56	40	23
8	53	36	23
9	51	35	24
10	49	34	25
11	47	32	26
12	45	32	27
13	43	32	28
14	41	32	29
15	40	32	30
16	39	28	31
17	38	28	32
18	37	28	33
19	36	28	34
20	35	28	35
21	34	26	36
22	33	26	37
23	32	26	38
24	31	26	39
25	30	26	40
26 - 30	30	24	15 kW + 1 kW
31 - 40	30	22	for each range
41 - 50	30	20	25 kW + 3/4 kW
51 - 60	30	18	for each range
61 and over	30	16	

See Next page for Notes to this table

DEMAND LOADS FOR HOUSEHOLD ELECTRIC RANGES, WALL-MOUNTED OVENS, COUNTER-MOUNTED COOKING UNITS, and OTHER HOUSEHOLD COOKING APPLIANCES over 1 3/4 kW RATING

NOTES

1. Over 12 kW through 27 kW ranges all of same rating. For ranges individually rated more than 12 kW but not more than 27 kW, the maximum demand in Column C shall be increased 5 percent for each additional kilowatt of rating or major fraction thereof by which the rating of individual ranges exceeds 12 kW.
2. Over 8¾ kW through 27 kW ranges of unequal ratings. For ranges individually rated more than 8¾ kW and of different ratings, but none exceeding 27 kW, an average value of rating shall be computed by adding together the ratings of all ranges to obtain the total connected load (using 12 kW for any range rated less than 12 kW) and dividing the total number of ranges. Then the maximum demand in Column C shall be increased 5 percent for each kilowatt or major fraction thereof by which this average value exceeds 12 kW.
3. Over 1¾ kW through 8¾ kW. In lieu of the method provided in Column C, it shall be permissible to add the nameplate ratings of all household cooking appliances rated more than 1¾ kW but not more than 8¾ kW and multiply the sum by the demand factors specified in Column A or B for the given number of appliances. Where the rating of cooking appliances falls under both Column A and Column B, the demand factors for each column shall be applied to the appliances for that column, and the results added together.
4. Branch-Circuit Load. It shall be permissible to compute the branch-circuit load for one range in accordance with Table 220.19. The branch-circuit load for one wall-mounted oven or one counter-mounted cooking unit shall be the nameplate rating of the appliance. The branch-circuit load for a counter-mounted cooking unit and not more than two wall-mounted ovens, all supplied from a single branch circuit and located in the same room, shall be computed by adding the nameplate rating of the individual appliances and treating this total as equivalent to one range.
5. This table also applies to household cooking appliances rated over 1¾ kW and used in instructional programs.

ELECTRICAL SYMBOLS

WALL	CEILING	
		OUTLET
D	D	DROP CORD
F	F	FAN OUTLET
J	J	JUNCTION BOX
L	L	LAMP HOLDER
L_{PS}	L_{PS}	LAMP HOLDER WITH PULL SWITCH
S	S	PULL SWITCH
V	V	VAPOR DISCHARGE SWITCH
X	X	EXIT OUTLET
C	C	CLOCK OUTLET
B	B	BLANKED OUTLET
		DUPLEX CONVENIENCE OUTLET
1,3		SINGLE, TRIPLEX, ETC.
		RANGE OUTLET
S		SWITCH AND CONVENIENCE OUTLET
		SPECIAL PURPOSE OUTLET
		FLOOR OUTLET

SWITCH OUTLETS	
S	SINGLE POLE SWITCH
S_2	DOUBLE POLE SWITCH
S_3	THREE WAY SWITCH
S_4	FOUR WAY SWITCH
S_D	AUTOMATIC DOOR SWITCH
S_E	ELECTROLIER SWITCH
S_P	SWITCH AND PILOT LAMP
S_K	KEY OPERATED SWITCH
S_{CB}	CIRCUIT BREAKER
S_{WCB}	WEATHER PROOF CIRCUIT BREAKER
S_{MC}	MOMENTARY CONTACT SWITCH
S_{RC}	REMOTE CONTROL SWITCH
S_{WP}	WEATHER PROOF SWITCH
S_F	FUSED SWITCH
S_{WPF}	WEATHER PROOF FUSED SWITCH
	LIGHTING SWITCH
	POWER PANEL

ELECTRICAL SYMBOLS

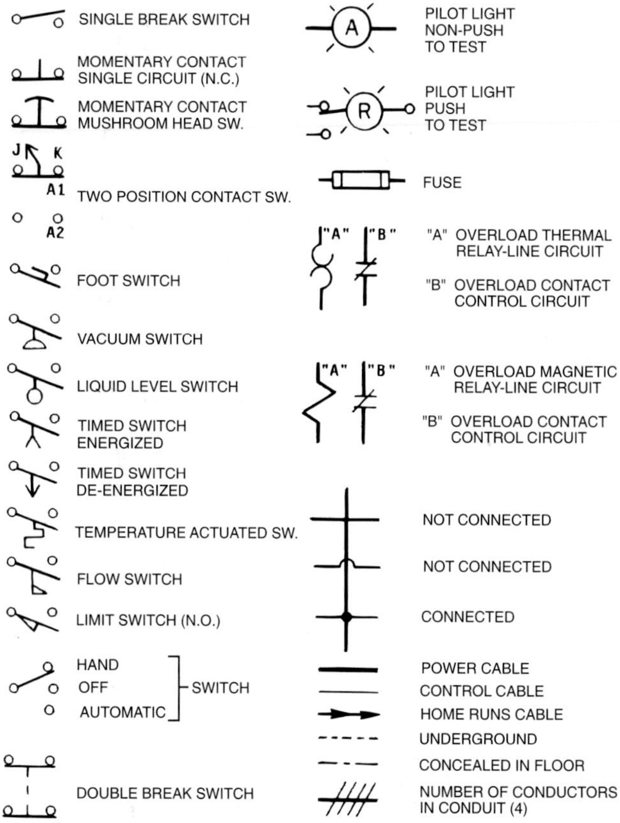

SINGLE BREAK SWITCH

MOMENTARY CONTACT
SINGLE CIRCUIT (N.C.)

MOMENTARY CONTACT
MUSHROOM HEAD SW.

J K
A1
TWO POSITION CONTACT SW.
A2

FOOT SWITCH

VACUUM SWITCH

LIQUID LEVEL SWITCH

TIMED SWITCH
ENERGIZED

TIMED SWITCH
DE-ENERGIZED

TEMPERATURE ACTUATED SW.

FLOW SWITCH

LIMIT SWITCH (N.O.)

HAND
OFF
AUTOMATIC
SWITCH

DOUBLE BREAK SWITCH

A
PILOT LIGHT
NON-PUSH
TO TEST

R
PILOT LIGHT
PUSH
TO TEST

FUSE

"A" "B"
"A" OVERLOAD THERMAL
RELAY-LINE CIRCUIT

"B" OVERLOAD CONTACT
CONTROL CIRCUIT

"A" "B"
"A" OVERLOAD MAGNETIC
RELAY-LINE CIRCUIT

"B" OVERLOAD CONTACT
CONTROL CIRCUIT

NOT CONNECTED

NOT CONNECTED

CONNECTED

POWER CABLE
CONTROL CABLE
HOME RUNS CABLE
UNDERGROUND
CONCEALED IN FLOOR
NUMBER OF CONDUCTORS
IN CONDUIT (4)

ELECTRICAL SYMBOLS

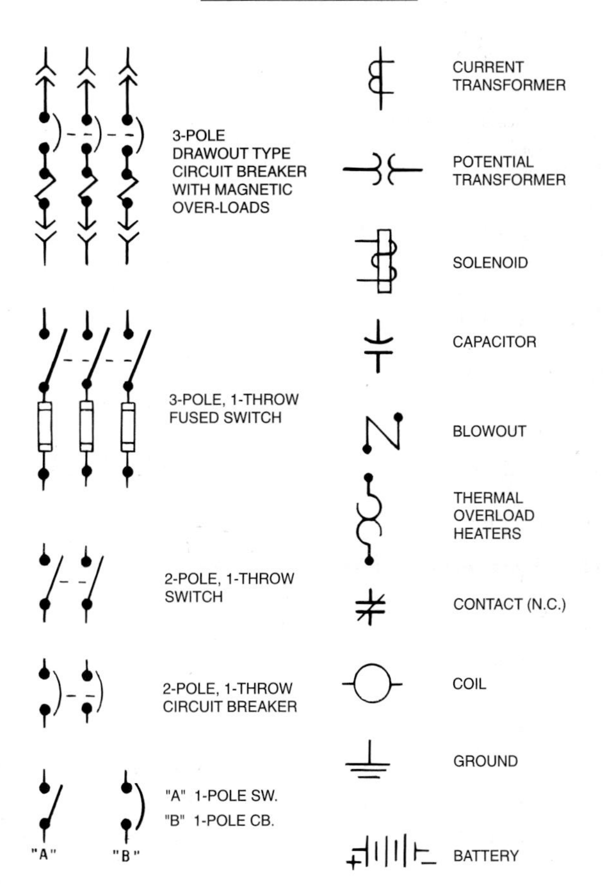

WIRING DIAGRAMS FOR NEMA CONFIGURATIONS

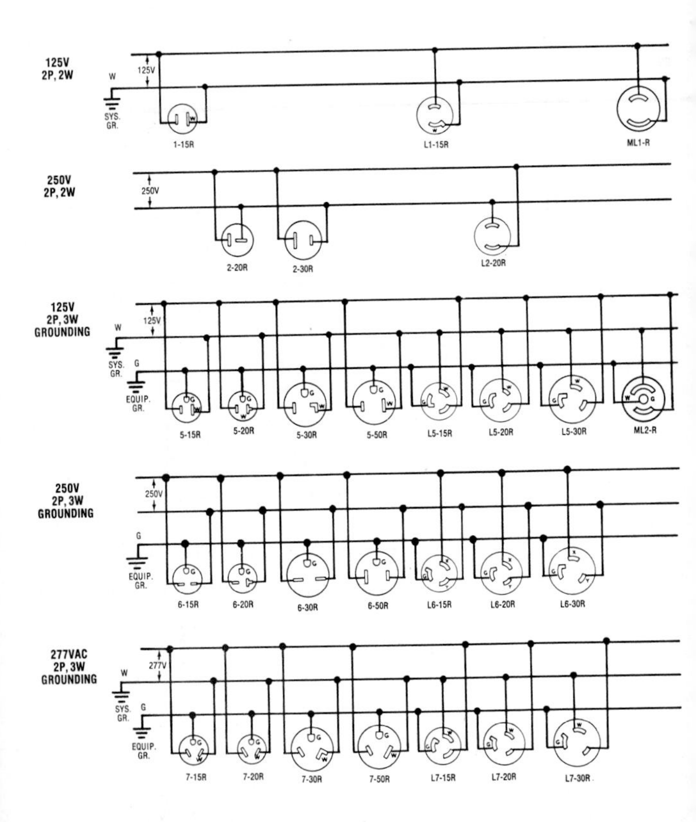

Courtesy of Cooper Industries, Inc. - Arrow Hart Wiring Devices

WIRING DIAGRAMS FOR NEMA CONFIGURATIONS

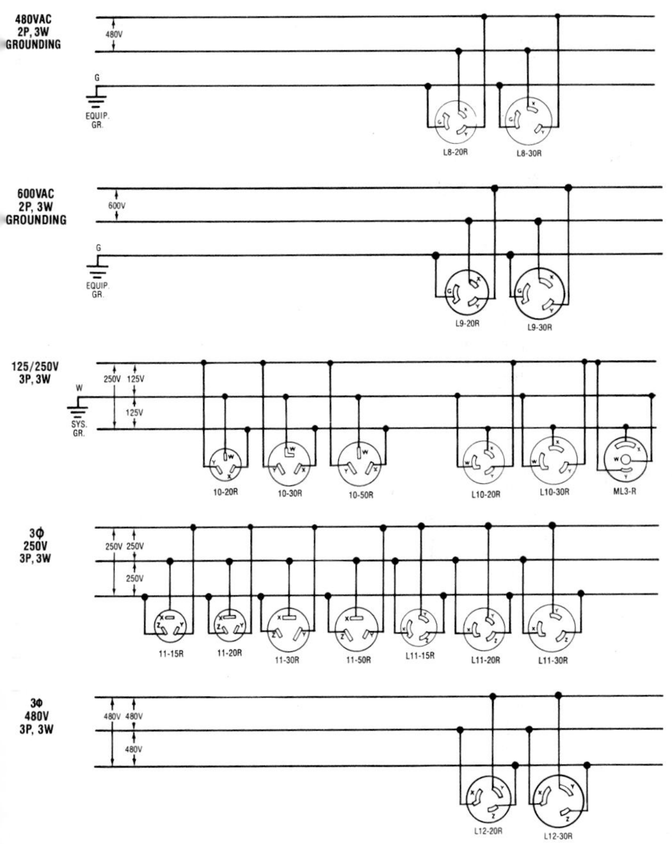

Courtesy of Cooper Industries, Inc. - Arrow Hart Wiring Devices

WIRING DIAGRAMS FOR NEMA CONFIGURATIONS

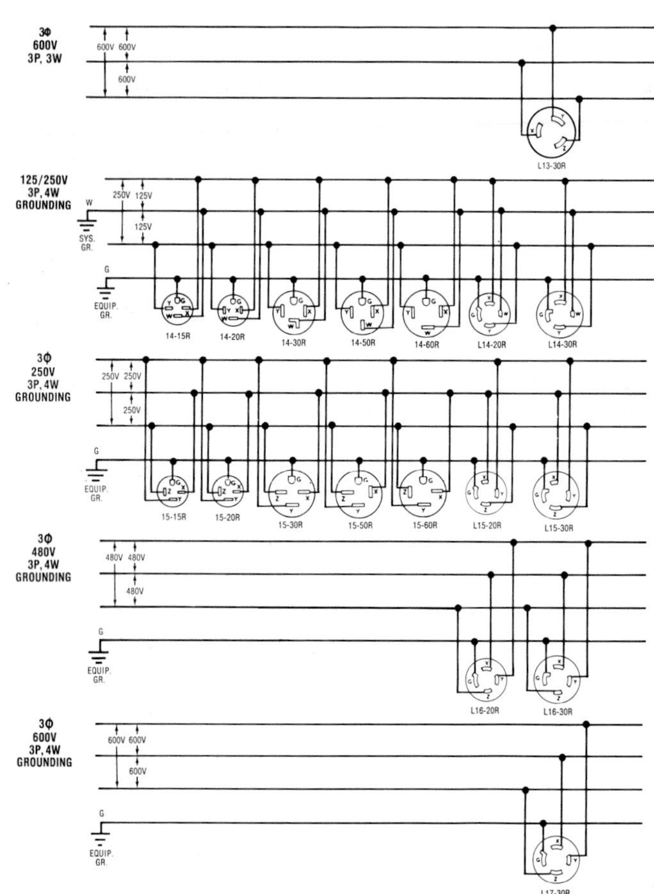

Courtesy of Cooper Industries, Inc. - Arrow Hart Wiring Devices

WIRING DIAGRAMS FOR NEMA CONFIGURATIONS

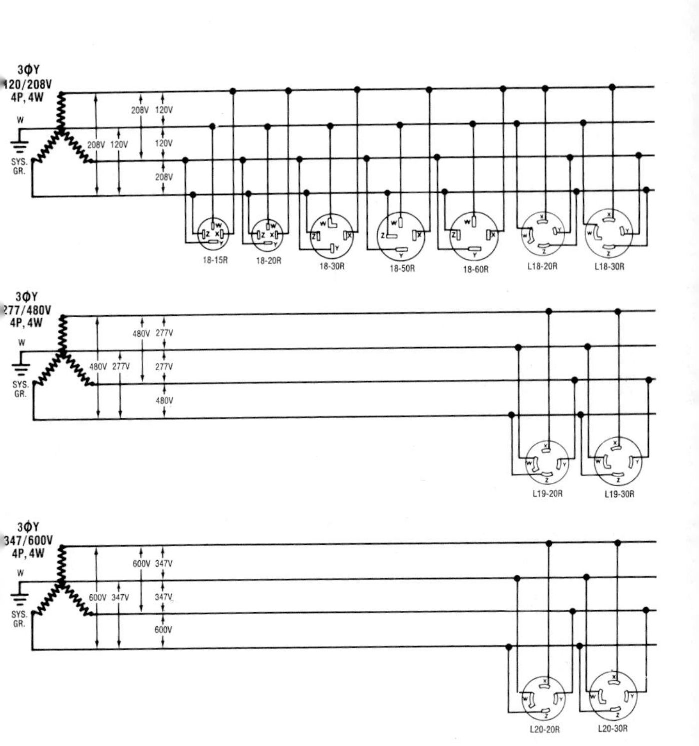

Courtesy of Cooper Industries, Inc. - Arrow Hart Wiring Devices

WIRING DIAGRAMS FOR NEMA CONFIGURATIONS

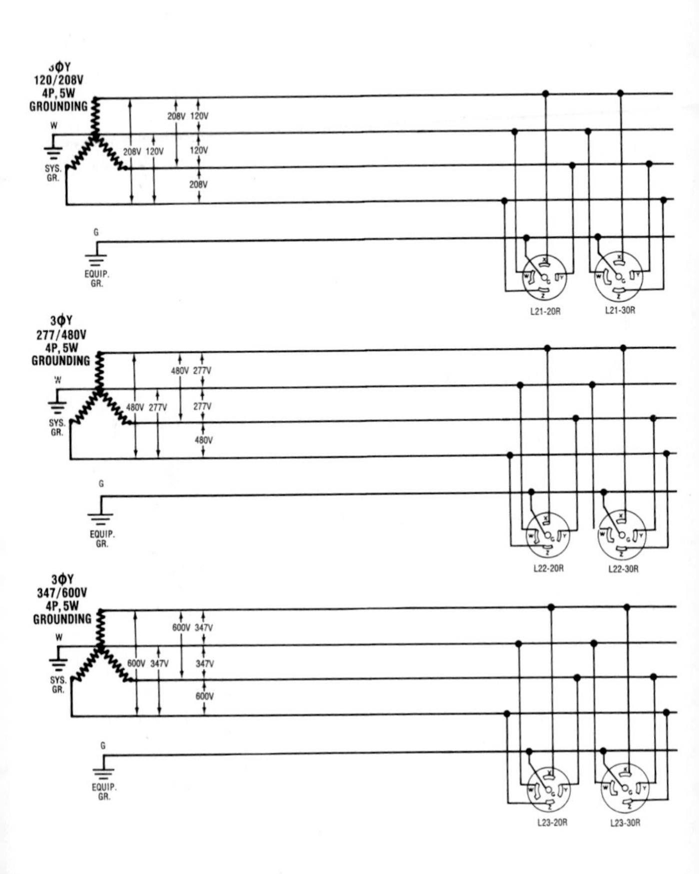

Courtesy of Cooper Industries, Inc. - Arrow Hart Wiring Devices

U.S. WEIGHTS AND MEASURES

LINEAR MEASURE

		1	INCH	=	2.540	CENTIMETERS
12	INCHES	= 1	FOOT	=	3.048	DECIMETERS
3	FEET	= 1	YARD	=	9.144	DECIMETERS
5.5	YARDS	= 1	ROD	=	5.029	METERS
40	RODS	= 1	FURLONG	=	2.018	HECTOMETERS
8	FURLONGS	= 1	MILE	=	1.609	KILOMETERS

MILE MEASUREMENTS

1	STATUTE MILE	=	5,280	FEET
1	SCOTS MILE	=	5,952	FEET
1	IRISH MILE	=	6,720	FEET
1	RUSSIAN VERST	=	3,504	FEET
1	ITALIAN MILE	=	4,401	FEET
1	SPANISH MILE	=	15,084	FEET

OTHER LINEAR MEASUREMENTS

1 HAND = 4 INCHES
1 SPAN = 9 INCHES
1 CHAIN = 22 YARDS
1 LINK = 7.92 INCHES
1 FATHOM = 6 FEET
1 FURLONG = 10 CHAINS
1 CABLE = 608 FEET

SQUARE MEASURE

144	SQUARE INCHES	=	1	SQUARE FOOT
9	SQUARE FEET	=	1	SQUARE YARD
30 1/4	SQUARE YARDS	=	1	SQUARE ROD
40	RODS	=	1	ROOD
4	ROODS	=	1	ACRE
640	ACRES	=	1	SQUARE MILE
1	SQUARE MILE	=	1	SECTION
36	SECTIONS	=	1	TOWNSHIP

CUBIC OR SOLID MEASURE

1	CU. FOOT	=	1728	CU. INCHES
1	CU. YARD	=	27	CU. FEET
1	CU. FOOT	=	7.48	GALLONS
1	GALLON (WATER)	=	8.34	LBS.
1	GALLON (U.S.)	=	231	CU. INCHES OF WATER
1	GALLON (IMPERIAL)	=	277 1/4	CU. INCHES OF WATER

U.S. WEIGHTS AND MEASURES

LIQUID MEASUREMENTS

1	PINT	=	4	GILLS
1	QUART	=	2	PINTS
1	GALLON	=	4	QUARTS
1	FIRKIN	=	9	GALLONS (ALE OR BEER)
1	BARREL	=	42	GALLONS (PETROLEUM OR CRUDE OIL)

DRY MEASURE

1	QUART	=	2	PINTS
1	PECK	=	8	QUARTS
1	BUSHEL	=	4	PECKS

WEIGHT MEASUREMENT (MASS)

A. AVOIRDUPOIS WEIGHT:

1	OUNCE	=	16	DRAMS
1	POUND	=	16	OUNCES
1	HUNDREDWEIGHT	=	100	POUNDS
1	TON	=	2,000	POUNDS

B. TROY WEIGHT:

1	CARAT	=	3.17	GRAINS
1	PENNYWEIGHT	=	20	GRAINS
1	OUNCE	=	20	PENNYWEIGHTS
1	POUND	=	12	OUNCES
1	LONG HUNDRED-WEIGHT	=	112	POUNDS
1	LONG TON	=	20	LONG HUNDREDWEIGHTS
		=	2240	POUNDS

C. APOTHECARIES WEIGHT:

1	SCRUPLE	=	20	GRAINS	=	1.296	GRAMS
1	DRAM	=	3	SCRUPLES	=	3.888	GRAMS
1	OUNCE	=	8	DRAMS	=	31.1035	GRAMS
1	POUND	=	12	OUNCES	=	373.2420	GRAMS

D. KITCHEN WEIGHTS AND MEASURES:

1	U.S. PINT	=	16	FL. OUNCES
1	STANDARD CUP	=	8	FL. OUNCES
1	TABLESPOON	=	0.5	FL. OUNCES (15 CU. CMS.)
1	TEASPOON	=	0.16	FL. OUNCES (5 CU. CMS.)

METRIC SYSTEM

PREFIXES:

A. MEGA	= 1,000,000		E. DECI	= 0.1	
B. KILO	= 1,000		F. CENTI	= 0.01	
C. HECTO	= 100		G. MILLI	= 0.001	
D. DEKA	= 10		H. MICRO	= 0.000001	

LINEAR MEASURE:

(THE UNIT IS THE METER = 39.37 INCHES)

1 CENTIMETER	= 10 MILLIMETERS	=	0.3937011	IN.
1 DECIMETER	= 10 CENTIMETERS	=	3.9370113	INS.
1 METER	= 10 DECIMETERS	=	1.0936143	YDS.
		=	3.2808429	FT.
1 DEKAMETER	= 10 METERS	=	10.936143	YDS.
1 HECTOMETER	= 10 DEKAMETERS	=	109.36143	YDS.
1 KILOMETER	= 10 HECTOMETERS	=	0.62137	MILE
1 MYRIAMETER	= 10,000 METERS			

SQUARE MEASURE:

(THE UNIT IS THE SQUARE METER = 1549.9969 SQ. INCHES)

1 SQ. CENTIMETER	= 100 SQ. MILLIMETERS	=	0.1550	SQ. IN.
1 SQ. DECIMETER	= 100 SQ. CENTIMETERS	=	15.550	SQ. INS.
1 SQ. METER	= 100 SQ. DECIMETERS	=	10.7639	SQ. FT.
1 SQ. DEKAMETER	= 100 SQ. METERS	=	119.60	SQ. YDS.
1 SQ. HECTOMETER	= 100 SQ. DEKAMETERS			
1 SQ. KILOMETER	= 100 SQ. HECTOMETERS			

(THE UNIT IS THE "ARE" = 100 SQ. METERS)

1 CENTIARE	= 10 MILLIARES	=	10.7643	SQ. FT.
1 DECIARE	= 10 CENTIARES	=	11.96033	SQ. YDS.
1 ARE	= 10 DECIARES	=	119.6033	SQ. YDS.
1 DEKARE	= 10 ARES	=	0.247110	ACRES
1 HEKTARE	= 10 DEKARES	=	2.471098	ACRES
1 SQ. KILOMETER	= 100 HEKTARES	=	0.38611	SQ. MILE

CUBIC MEASURE:

(THE UNIT IS THE "STERE" = 61,025.38659 CU. INS.)

1 DECISTERE	= 10 CENTISTERES	=	3.531562	CU. FT.
1 STERE	= 10 DECISTERES	=	1.307986	CU. YDS.
1 DEKASTERE	= 10 STERES	=	13.07986	CU. YDS.

METRIC SYSTEM

CUBIC MEASURE:

(THE UNIT IS THE METER = 39.37 INCHES)

1 CU. CENTIMETER	= 1000 CU. MILLIMETERS	=	0.06102 CU. IN.
1 CU. DECIMETER	= 1000 CU. CENTIMETERS	=	61.02374 CU. IN.
1 CU. METER	= 1000 CU. DECIMETERS	=	35.31467 CU. FT.
	= 1 STERE	=	1.30795 CU. YDS.
1 CU. CENTIMETER (WATER)		=	1 GRAM
1000 CU. CENTIMETERS (WATER)	= 1 LITER	=	1 KILOGRAM
1 CU. METER (1000 LITERS)		=	1 METRIC TON

MEASURES OF WEIGHT:

(THE UNIT IS THE GRAM = 0.035274 OUNCES)

1 MILLIGRAM	=	=	0.015432	GRAINS
1 CENTIGRAM	= 10 MILLIGRAMS	=	0.15432	GRAINS
1 DECIGRAM	= 10 CENTIGRAMS	=	1.5432	GRAINS
1 GRAM	= 10 DECIGRAMS	=	15.4323	GRAINS
1 DEKAGRAM	= 10 GRAMS	=	5.6438	DRAMS
1 HECTOGRAM	= 10 DEKAGRAMS	=	3.5274	OUNCES
1 KILOGRAM	= 10 HECTOGRAMS	=	2.2046223	POUNDS
1 MYRIAGRAM	= 10 KILOGRAMS	=	22.046223	POUNDS
1 QUINTAL	= 10 MYRIAGRAMS	=	1.986412	CWT.
1 METRIC TON	= 10 QUINTAL	=	2,2045.622	POUNDS
1 GRAM	= 0.56438 DRAMS			
1 DRAM	= 1.77186 GRAMS			
	= 27.3438 GRAINS			
1 METRIC TON	= 2,204.6223 POUNDS			

MEASURES OF CAPACITY:

(THE UNIT IS THE LITER = 1.0567 LIQUID QUARTS)

1 CENTILITER	= 10 MILLILITERS	=	0.338	FLUID OUNCES
1 DECILITER	= 10 CENTILITERS	=	3.38	FLUID OUNCES
1 LITER	= 10 DECILITERS	=	33.8	FLUID OUNCES
1 DEKALITER	= 10 LITERS	=	0.284	BUSHEL
1 HECTOLITER	= 10 DEKALITERS	=	2.84	BUSHELS
1 KILOLITER	= 10 HECTOLITERS	=	264.2	GALLONS

NOTE: $\frac{\text{KILOMETERS}}{8} \times 5 = \text{MILES}$ or $\frac{\text{MILES}}{5} \times 8 = \text{KILOMETERS}$

METRIC DESIGNATOR AND TRADE SIZES

METRIC DESIGNATOR												
12	16	21	27	35	41	53	63	78	91	103	129	155
3/8	1/2	3/4	1	1 1/4	1 1/2	2	2 1/2	3	3 1/2	4	5	6
TRADE SIZE												

U.S. WEIGHTS & MEASURES / METRIC EQUIVALENT CHART

	In.	Ft.	Yd.	Mile	Mm	Cm	M	Km
1 Inch =	1	.0833	.0278	1.578×10^{-5}	25.4	**2.54**	.0254	2.54×10^{-5}
1 Foot =	12	1	.333	1.894×10^{-4}	304.8	**30.48**	.3048	3.048×10^{-4}
1 Yard =	36	3	1	5.6818×10^{-4}	914.4	91.44	**.9144**	9.144×10^{-4}
1 Mile =	63,360	5,280	1,760	1	1,609,344	160,934.4	1,609.344	**1.609344**
1 mm =	**.03937**	.0032808	1.0936×10^{-3}	6.2137×10^{-7}	1	0.1	0.001	0.000001
1 cm =	**.3937**	.0328084	.0109361	6.2137×10^{-6}	10	1	0.01	0.00001
1 m =	39.37	3.28084	**1.09361**	6.2137×10^{-4}	1000	100	1	0.001
1 km =	39,370	3,280.84	1,093.61	**0.62137**	1,000,000	100,000	1,000	1

In. = Inches Ft. = Foot Yd. = Yard Mi. = Mile Mm = Millimeter Cm = Centimeter M = Meter Km = Kilometer

EXPLANATION OF SCIENTIFIC NOTATION:

Scientific Notation is simply a way of expressing very large or very small numbers in a more compact format. Any number can be expressed as a number between 1 & 10, multiplied by a power of 10 (which indicates the correct position of the decimal point in the original number). Numbers greater than 10 have positive powers of 10, and numbers less than 1 have negative powers of 10.

Example: $186{,}000 = 1.86 \times 10^5$ $0.000524 = 5.24 \times 10^{-4}$

USEFUL CONVERSIONS / EQUIVALENTS

1 BTU .... Raises 1 LB. of water 1°F
1 GRAM CALORIE ..... Raises 1 Gram of water 1°C
1 CIRCULAR MIL ..... Equals 0.7854 sq. mil
1 SQ. MIL .... Equals 1.27 cir. mils
1 MIL .... Equals 0.001 in.

To determine circular mil of a conductor:

ROUND CONDUCTOR .... $CM = (\text{Diameter in mils})^2$

BUS BAR .... $CM = \dfrac{\text{Width (mils)} \times \text{Thickness (mils)}}{0.7854}$

NOTES: 1 Millimeter = 39.37 Mils   1 Cir. Millimeter = 1550 Cir. Mils
1 Sq. Millimeter = 1974 Cir. Mils

DECIMAL EQUIVALENTS

FRACTION					DECIMAL
1/64					.0156
2/64	1/32				.0313
3/64					.0469
4/64	2/32	1/16			.0625
5/64					.0781
6/64	3/32				.0938
7/64					.1094
8/64	4/32	2/16	1/8		.125
9/64					.1406
10/64	5/32				.1563
11/64					.1719
12/64	6/32	3/16			.1875
13/64					.2031
14/64	7/32				.2188
15/64					.2344
16/64	8/32	4/16	2/8	1/4	.25
17/64					.2656
18/64	9/32				.2813
19/64					.2969
20/64	10/32	5/16			.3125
21/64					.3281
22/64	11/32				.3438
23/64					.3594
24/64	12/32	6/16	3/8		.375
25/64					.3906
26/64	13/32				.4063
27/64					.4219
28/64	14/32	7/16			.4375
29/64					.4531
30/64	15/32				.4688
31/64					.4844
32/64	16/32	8/16	4/8	2/4	.5
33/64					.5156
34/64	17/32				.5313
35/64					.5469
36/64	18/32	9/16			.5625
37/64					.5781
38/64	19/32				.5938
39/64					.6094
40/64	20/32	10/16	5/8		.625
41/64					.6406
42/64	21/32				.6563
43/64					.6719
44/64	22/32	11/16			.6875
45/64					.7031
46/64	23/32				.7188
47/64					.7344
48/64	24/32	12/16	6/8	3/4	.75
49/64					.7656
50/64	25/32				.7813
51/64					.7969
52/64	26/32	13/16			.8125
53/64					.8281
54/64	27/32				.8438
55/64					.8594
56/64	28/32	14/16	7/8		.875
57/64					.8906
58/64	29/32				.9063
59/64					.9219
60/64	30/32	15/16			.9375
61/64					.9531
62/64	31/32				.9688
63/64					.9844
64/64	32/32	16/16	8/8	4/4	1.000

Decimals are rounded to the nearest 10,000th.

TWO-WAY CONVERSION TABLE

To convert from the unit of measure in Column B to the unit of measure in Column C, multiply the number of units in Column B by the multiplier in Column A. To convert from Column C to B, use the multiplier in Column D.

EXAMPLE: To convert 1000 BTU's to CALORIES, find the "BTU - CALORIE" combination in Columns B and C. "BTU" is in Column B and "CALORIE" is in Column C; so we are converting from B to C. Therefore, we use Column A multiplier. 1000 BTU's x 251.996 = 251,996 Calories.

To convert 251,996 Calories to BTU's, use the same "BTU - CALORIE" combination. But this time you are converting from C to B. Therefore, use Column D multiplier. 251,996 Calories x .0039683 = 1,000 BTU's.

A x B = C & **D x C = B**

To convert from B to C, Multiply by A:

To convert from C to B, Multiply by D:

A	B	C	D
43,560	Acre	Sq. Foot	2.2956×10^{-5}
1.5625×10^{-3}	Acre	Sq. Mile	640
6.4516	Ampere per sq. cm.	Ampere per sq. in.	.155003
1.256637	Ampere (turn)	Gilberts	0.79578
33.89854	Atmosphere	Foot of H_2O	0.029499
29.92125	Atmosphere	Inch of Hg	0.033421
14.69595	Atmosphere	Pound force/sq. in.	0.06804
251.996	BTU	Calorie	3.96832×10^{-3}
778.169	BTU	Foot-pound force	1.28507×10^{-3}
3.93015×10^{-4}	BTU	Horsepower-hour	2544.43
1055.056	BTU	Joule	9.47817×10^{-4}
2.9307×10^{-4}	BTU	Kilowatt-hour	3412.14
3.93015×10^{-4}	BTU/hour	Horsepower	2544.43
2.93071×10^{-4}	BTU/hour	Kilowatt	3412.1412
0.293071	BTU/hour	Watt	3.41214
4.19993	BTU/minute	Calorie/second	0.23809
0.0235809	BTU/minute	Horsepower	42.4072
17.5843	BTU/minute	Watt	0.0568

TWO-WAY CONVERSION TABLE (continued)

To convert from B to C, Multiply by A:			To convert from C to B, Multiply by D:
A	**B**	**C**	**D**
4.1868	Calorie	Joule	.238846
0.0328084	Centimeter	Foot	30.48
0.3937	Centimeter	Inch	2.54
0.00001	Centimeter	Kilometer	100,000
0.01	Centimeter	Meter	100
6.2137×10^{-6}	Centimeter	Mile	160,934.4
10	Centimeter	Millimeter	0.1
0.010936	Centimeter	Yard	91.44
7.85398×10^{-7}	Circular mil	Sq. Inch	1.273239×10^{6}
0.000507	Circular mil	Sq. Millimeter	1973.525
0.06102374	Cubic Centimeter	Cubic Inch	16.387065
0.028317	Cubic Foot	Cubic Meter	35.31467
1.0197×10^{-3}	Dyne	Gram Force	980.665
1×10^{-5}	Dyne	Newton	100,000
1	Dyne centimeter	Erg	1
7.376×10^{-8}	Erg	Foot pound force	1.355818×10^{7}
2.777×10^{-14}	Erg	Kilowatt-hour	3.6×10^{13}
1.0×10^{-7}	Erg/second	Watt	1.0×10^{7}
12	Foot	Inch	0.0833
3.048×10^{-4}	Foot	Kilometer	3,280.84
0.3048	Foot	Meter	3.28084
1.894×10^{-4}	Foot	Mile	5,280
304.8	Foot	Millimeter	0.00328
0.333	Foot	Yard	3
10.7639	Foot candle	Lux	0.0929
0.882671	Foot of H_2O	Inch of Hg	1.13292
5.0505×10^{-7}	Foot pound force	Horsepower-hour	1.98×10^{6}
1.35582	Foot pound force	Joule	0.737562
3.76616×10^{-7}	Foot pound force	Kilowatt-hour	2.655223×10^{6}
3.76616×10^{-4}	Foot pound force	Watt-hour	2655.22
3.76616×10^{-7}	Foot pnd force/hour	Kilowatt	2.6552×10^{6}
3.0303×10^{-5}	Foot pnd force/minute	Horsepower	33,000

TWO-WAY CONVERSION TABLE (continued)

To convert from B to C, Multiply by A: | To convert from C to B, Multiply by D:

A	B	C	D
2.2597×10^{-5}	Foot pnd force/minute	Kilowatt	44,253.7
0.022597	Foot pnd force/minute	Watt	44.2537
1.81818×10^{-3}	Foot pnd force/second	Horsepower	550
1.355818×10^{-3}	Foot pnd force/second	Kilowatt	737.562
0.7457	Horsepower	Kilowatt	1.34102
745.7	Horsepower	Watt	0.00134
.0022046	Gram	Pound mass	453.592
2.54×10^{-5}	Inch	Kilometer	39,370
0.0254	Inch	Meter	39.37
1.578×10^{-5}	Inch	Mile	63,360
25.4	Inch	Millimeter	0.03937
0.0278	Inch	Yard	36
0.07355	Inch of H_2O	Inch of Hg	13.5951
2.7777×10^{-7}	Joule	Kilowatt-hour	3.6×10^{6}
2.7777×10^{-4}	Joule	Watt hour	3600
1	Joule	Watt second	1
1,000	Kilometer	Meter	0.001
0.62137	Kilometer	Mile	1.609344
1,000,000	Kilometer	Millimeter	0.000001
1,093.61	Kilometer	Yard	9.144×10^{-4}
0.000621	Meter	Mile	1,609.344
1,000	Meter	Millimeter	0.001
1.0936	Meter	Yard	0.9144
1,609,344	Mile	Millimeter	6.2137×10^{-7}
1,760	Mile	Yard	5.6818×10^{-4}
1.0936×10^{-3}	Millimeter	Yard	914.4
0.224809	Newton	Pound force	4.44822
0.03108	Pound	Slug	32.174
0.0005	Pound	Ton (short)	2,000
0.155	Sq. Centimeter	Sq. Inch	6.4516
0.092903	Sq. Foot	Sq. Meter	10.76391
0.386102	Sq. Kilometer	Sq. Mile	2.589988

METALS

METAL	SYMB	SPEC. GRAV.	MELT POINT C°	MELT POINT F°	ELEC. COND. % COPPER	LBS. CU. "
ALUMINUM	AL	2.71	660	1220	64.9	.0978
ANTIMONY	SB	6.62	630	1167	4.42	.2390
ARSENIC	AS	5.73	-	-	4.9	.2070
BERYLLIUM	BE	1.83	1280	2336	9.32	.0660
BISMUTH	BI	9.80	271	520	1.50	.3540
BRASS (70-30)		8.51	900	1652	28.0	.3070
BRONZE (5% SN)		8.87	1000	1382	18.0	.3200
CADMIUM	CD	8.65	321	610	22.7	.3120
CALCIUM	CA	1.55	850	1562	50.1	.0560
COBALT	CO	8.90	1495	2723	17.8	.3210
COPPER	CU					
ROLLED		8.89	1083	1981	100.0	.3210
TUBING		8.95	-	-	100.0	.3230
GOLD	AU	19.30	1063	1945	71.2	.6970
GRAPHITE		2.25	3500	6332	10^{-3}	.0812
INDIUM	IN	7.30	156	311	20.6	.2640
IRIDIUM	IR	22.40	2450	4442	32.5	.8090
IRON	FE	7.20	1200 TO 1400	2192 TO 2552	17.6	.2600
MALLEABLE		7.20	1500 TO 1600	2732 TO 2912	10	.2600
WROUGHT		7.70	1500 TO 1600	2732 TO 2912	10	.2780
LEAD	PB	11.40	327	621	8.35	.4120
MAGNESIUM	MG	1.74	651	1204	38.7	.0628
MANGANESE	MN	7.20	1245	2273	0.9	.2600
MERCURY	HG	13.65	-38.9	-37.7	1.80	.4930
MOLYBDENUM	MO	10.20	2620	4748	36.1	.3680
MONEL (63 - 37)		8.87	1300	2372	3.0	.3200
NICKEL	NI	8.90	1452	2646	25.0	.3210
PHOSPHOROUS	P	1.82	44.1	111.4	10^{-17}	.0657
PLATINUM	PT	21.46	1773	3221	17.5	.7750
POTASSIUM	K	0.860	62.3	144.1	28	.0310
SELENIUM	SE	4.81	220	428	14.4	.1740
SILICON	SI	2.40	1420	2588	10^{-5}	.0866
SILVER	AG	10.50	960	1760	106	.3790
STEEL (CARBON)		7.84	1330 TO 1380	2436 TO 2516	10	.2830
STAINLESS						
(18-8)		7.92	1500	2732	2.5	.2860
(13-CR)		7.78	1520	2768	3.5	.2810

METALS

METAL	SYMB	SPEC. GRAV.	MELT POINT C°	MELT POINT F°	ELEC. COND. % COPPER	LBS. CU. "
TANTALUM	TA	16.60	2900	5414	13.9	.599
TELLURIUM	TE	6.20	450	846	10^{-5}	.224
THORIUM	TH	11.70	1845	3353	9.10	.422
TIN	SN	7.30	232	449	15.00	.264
TITANIUM	TI	4.50	1800	3272	2.10	.162
TUNGSTEN	W	19.30	3410	-	31.50	.697
URANIUM	U	18.70	1130	2066	2.80	.675
VANADIUM	V	5.96	1710	3110	6.63	.215
ZINC	ZN	7.14	419	786	29.10	.258
ZIRCONIUM	ZR	6.40	1700	3092	4.20	.231

SPECIFIC RESISTANCE (K)

THE SPECIFIC RESISTANCE (K) OF A MATERIAL IS THE RESISTANCE OFFERED BY A WIRE OF THIS MATERIAL WHICH IS ONE FOOT LONG WITH A DIAMETER OF 1 MIL.

MATERIAL	"K"	MATERIAL	"K"
BRASS	43.0	ALUMINUM	17.0
CONSTANTAN	295	MONEL	253
COPPER	10.8	NICHROME	600
GERMAN SILVER 18%	200	NICKEL	947
GOLD	14.7	TANTALUM	93.3
IRON (PURE)	60.0	TIN	69.0
MAGNESIUM	276	TUNGSTEN	34.0
MANGANIN	265	SILVER	9.7

NOTE:
1. The resistance of a wire is directly proportional to the specific resistance of the material.
2. "K" = Specific Resistance
3. Resistance varies with temperature. See NEC, Chapter 9, Table 8, NOTES

CENTIGRADE AND FAHRENHEIT THERMOMETER SCALES

DEG-C	DEG-F	DEG-C	DEG-F	DEG-C	DEG-F	DEG-C	DEG-F
0	32						
1	33.8	26	78.8	51	123.8	76	168.8
2	35.6	27	80.6	52	125.6	77	170.6
3	37.4	28	82.4	53	127.4	78	172.4
4	39.2	29	84.2	54	129.2	79	174.2
5	41	30	86	55	131	80	176
6	42.8	31	87.8	56	132.8	81	177.8
7	44.6	32	89.6	57	134.6	82	179.6
8	46.4	33	91.4	58	136.4	83	181.4
9	48.2	34	93.2	59	138.2	84	183.2
10	50	35	95	60	140	85	185
11	51.8	36	96.8	61	141.8	86	186.8
12	53.6	37	98.6	62	143.6	87	188.6
13	55.4	38	100.4	63	145.4	88	190.4
14	57.2	39	102.2	64	147.2	89	192.2
15	59	40	104	65	149	90	194
16	60.8	41	105.8	66	150.8	91	195.8
17	62.6	42	107.6	67	152.6	92	197.6
18	64.4	43	109.4	68	154.4	93	199.4
19	66.2	44	111.2	69	156.2	94	201.2
20	68	45	113	70	158	95	203
21	69.8	46	114.8	71	159.8	96	204.8
22	71.6	47	116.6	72	161.6	97	206.6
23	73.4	48	118.4	73	163.4	98	208.4
24	75.2	49	120.2	74	165.2	99	210.2
25	77	50	122	75	167	100	212

1. TEMP. C° = 5/9 x (TEMP. F° - 32)
2. TEMP F° = (9/5 x TEMP. C°) + 32
3. Ambient temperature is the temperature of the surrounding cooling medium.
4. Rated temperature rise is the permissible rise in temperature above ambient when operating under load.

USEFUL MATH FORMULAS

RIGHT TRIANGLE

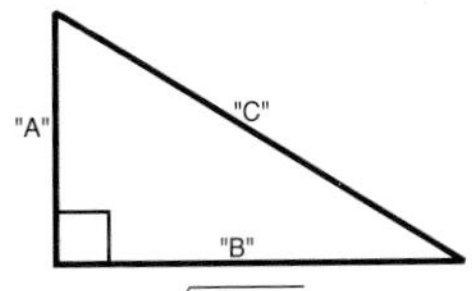

$A = \sqrt{C^2 - B^2}$

$B = \sqrt{C^2 - A^2}$

$C = \sqrt{A^2 + B^2}$

OBLIQUE TRIANGLE

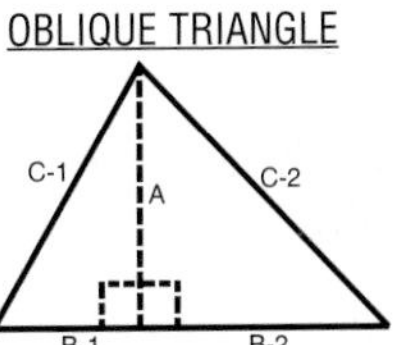

SOLVE AS TWO RIGHT TRIANGLES

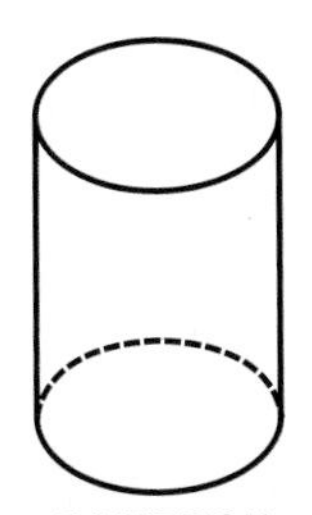

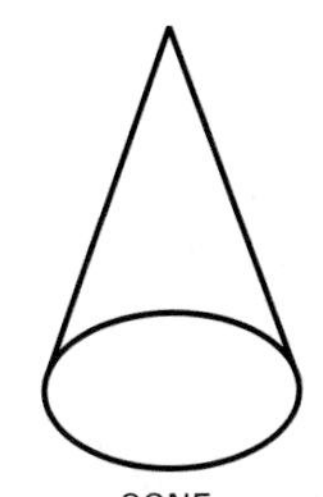

SPHERE

AREA = D^2 X 3.1416
VOLUME = D^3 X 0.5236

CYLINDRICAL

VOLUME =
AREA OF END X HEIGHT

CONE

VOLUME =
AREA OF END X HEIGHT ÷ 3

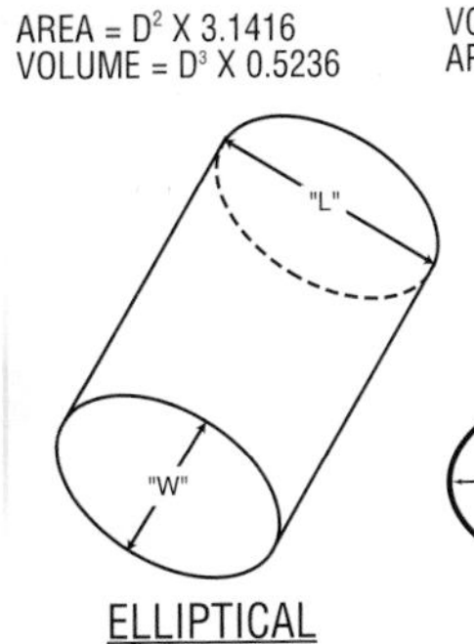

ELLIPTICAL

SOLVE THE SAME AS CYLINDRICAL

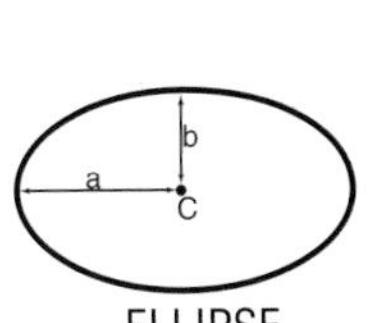

ELLIPSE

AREA = a x b x 3.1416
(assuming C is center)

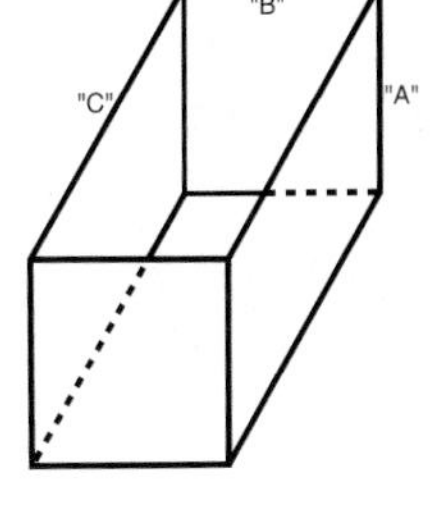

VOLUME = A x B x C

See next page for CIRCLE

THE CIRCLE

DEFINITION: A closed plane curve having every point an equal distance from a fixed point within the curve.

CIRCUMFERENCE : The distance around a circle
DIAMETER : The distance across a circle through the center
RADIUS : The distance from the center to the edge of a circle
ARC : A part of the circumference
CHORD : A straight line connecting the ends of an arc.
SEGMENT : An area bounded by an arc and a chord
SECTOR : A part of a circle enclosed by two radii and the arc which they cut off

CIRCUMFERENCE OF A CIRCLE = 3.1416 x 2 x Radius
AREA OF A CIRCLE = 3.1416 x Radius^2
ARC LENGTH = Degrees in arc x radius x 0.01745
RADIUS LENGTH = one half length of diameter
SECTOR AREA = one half length of arc x radius
CHORD LENGTH = $2\sqrt{A \times B}$
SEGMENT AREA = Sector area minus triangle area

NOTE:
3.1416 x 2 x R = 360 Degrees,
or 0.0087266 x 2 x R = 1 Degree,
or 0.01745 x R = 1 Degree
This gives us the arc formula.
DEGREES x RADIUS x 0.01745 = DEVELOPED LENGTH

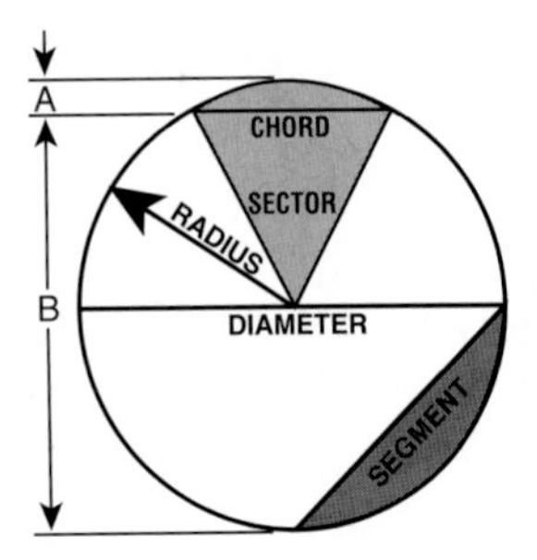

EXAMPLE:
For a ninety degree conduit bend, having a radius of 17.25":
90 x 17.25" x 0.01745 = Developed Length
27.09" = Developed Length

FRACTIONS

DEFINITIONS:

A. A FRACTION is a quantity less than a unit.

B. A NUMERATOR is the term of a fraction indicating how many of the parts of a unit are to be taken. In a common fraction, it appears above or to the left of the line.

C. A DENOMINATOR is the term of a fraction indicating the number of equal parts into which the unit is divided. In a common fraction, it appears below or to the right of the line.

D. EXAMPLES:

(1.) $\frac{\mathbf{1}}{\mathbf{2}} = \frac{\text{NUMERATOR}}{\text{DENOMINATOR}} = \text{FRACTION}$ (1 → NUMERATOR; 2 → DENOMINATOR)

(2.) NUMERATOR ⟶ **1/2** ⟵ DENOMINATOR

TO ADD OR SUBTRACT:

TO SOLVE: 1/2 - 2/3 + 3/4 - 5/6 + 7/12 = ?

A. Determine the lowest common denominator that each of the denominators 2, 3, 4, 6, and 12 will divide into an even number of times.
The lowest common denominator is 12.

B. Work one fraction at a time using the formula:

$$\frac{\textbf{COMMON DENOMINATOR}}{\textbf{DENOMINATOR OF FRACTION}} \times \textbf{NUMERATOR OF FRACTION}$$

(1.) 12/2 x 1 = 6 x 1 = 6 — 1/2 becomes 6/12
(2.) 12/3 x 2 = 4 x 2 = 8 — 2/3 becomes 8/12
(3.) 12/4 x 3 = 3 x 3 = 9 — 3/4 becomes 9/12
(4.) 12/6 x 5 = 2 x 5 = 10 — 5/6 becomes 10/12
(5.) 7/12 remains 7/12

(continued next page)

TO ADD OR SUBTRACT *(CONTINUED)*:

C. We can now convert the problem from its original form to its new form using 12 as the common denominator.

$1/2 - 2/3 + 3/4 - 5/6 + 7/12 =$ Original form

$\frac{6 - 8 + 9 - 10 + 7}{12} =$ Present form

$\frac{4}{12} = \frac{1}{3}$ Reduced to lowest form

D. To convert fractions to decimal form, simply divide the numerator of the fraction by the denominator of the fraction.

EXAMPLE: $\frac{1}{3} =$ 1 DIVIDED BY 3 = 0.333

TO MULTIPLY:

A. The numerator of fraction #1 times the numerator of fraction #2 is equal to the numerator of the product.

B. The denominator of fraction #1 times the denominator of fraction #2 is equal to the denominator of the product.

C. EXAMPLE:

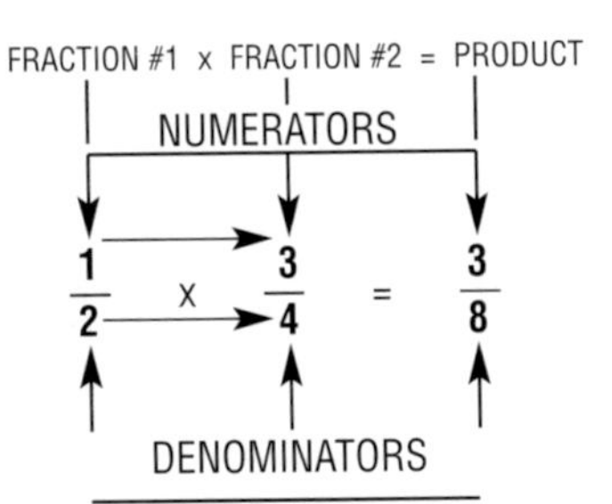

NOTE: To change 3/8 to decimal form, divide 3 by 8 = .375

FRACTIONS

TO DIVIDE:

A. The numerator of fraction #1 times the denominator of fraction #2 is equal to the numerator of the quotient.

B. The denominator of fraction #1 times the numerator of fraction #2 is equal to the denominator of the quotient.

C. EXAMPLE: $\frac{1}{2} \div \frac{3}{4}$

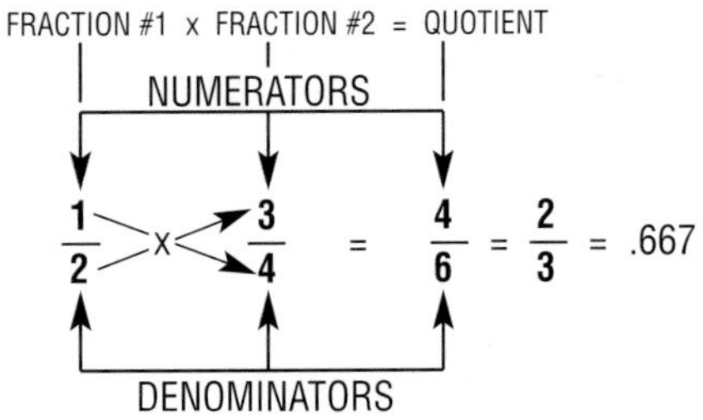

FRACTION #1 x FRACTION #2 = QUOTIENT

NUMERATORS

$$\frac{\mathbf{1}}{\mathbf{2}} \times \frac{\mathbf{3}}{\mathbf{4}} = \frac{\mathbf{4}}{\mathbf{6}} = \frac{\mathbf{2}}{\mathbf{3}} = .667$$

DENOMINATORS

D. Alternate method for dividing by a fraction is to multiply by the reciprocal of the divisor. (the second fraction in a division problem).

E. EXAMPLE: $\frac{1}{2} \div \frac{3}{4}$

The reciprocal of $\frac{3}{4}$ is $\frac{4}{3}$

so, $\frac{1}{2} \div \frac{3}{4} = \frac{1}{2} \times \frac{4}{3} = \frac{4}{6} = \frac{2}{3} = .667$

EQUATIONS

The word "EQUATION" means equal or the same as.

A. EXAMPLE: $2 \times 10 = 4 \times 5$

$20 = 20$

RULES:

A. **The same number may be added to both sides of an equation without changing its values.**

EXAMPLE: $(2 \times 10) + 3 = (4 \times 5) + 3$

$23 = 23$

B. **The same number may be subtracted from both sides of an equation without changing its values.**

EXAMPLE: $(2 \times 10) - 3 = (4 \times 5) - 3$

$17 = 17$

C. **Both sides of an equation may be divided by the same number without changing its values.**

EXAMPLE: $\frac{2 \times 10}{20} = \frac{4 \times 5}{20}$

$1 = 1$

D. **Both sides of an equation may be multiplied by the same number without changing its values.**

EXAMPLE: $3 \times (2 \times 10) = 3 \times (4 \times 5)$

$60 = 60$

E. **TRANSPOSITION:**

The process of moving a quantity from one side of an equation to the other side of an equation by changing its sign of operation.

1. **A term may be transposed if its sign is changed from plus (+) to minus (-), or from minus (-) to plus (+).**

EXAMPLE: $X + 5 = 25$

$X + 5 - 5 = 25 - 5$

$X = 20$

E. **TRANSPOSITION** (continued):

2. **A multiplier may be removed from one side of an equation by making it a divisor in the other side; or a divisor may be removed from one side of an equation by making it a multiplier in the other side.**

EXAMPLE: Multiplier from one side of equation (4) becomes divisor in other side.

$$4X = 40 \quad \text{becomes} \quad X = \frac{40}{4} = 10$$

EXAMPLE: Divisor from one side of equation becomes multiplier in other side.

$$\frac{X}{4} = 10 \quad \text{becomes} \quad X = 10 \times 4$$

SIGNS:

A. **ADDITION** of numbers with *DIFFERENT* signs:

1. **RULE: Use the sign of the larger and subtract.**

EXAMPLE:

$$\begin{array}{r} +3 \\ +\ -2 \\ \hline +1 \end{array} \qquad \begin{array}{r} -2 \\ +\ +3 \\ \hline +1 \end{array}$$

B. **ADDITION** of numbers with the *SAME* signs:

2. **RULE: Use the common sign and add.**

EXAMPLE:

$$\begin{array}{r} +3 \\ +\ +2 \\ \hline +5 \end{array} \qquad \begin{array}{r} -3 \\ +\ -2 \\ \hline -5 \end{array}$$

C. **SUBTRACTION** of numbers with *DIFFERENT* signs:

3. **RULE: Change the sign of the subtrahend (the second number in a subtraction problem) and proceed as in addition.**

EXAMPLE:

$$\begin{array}{r} +3 \\ -\ -2 \\ \hline \end{array} = \begin{array}{r} +3 \\ +\ +2 \\ \hline +5 \end{array} \qquad \begin{array}{r} -2 \\ -\ +3 \\ \hline \end{array} = \begin{array}{r} -2 \\ +\ -3 \\ \hline -5 \end{array}$$

SIGNS (continued):

D. **SUBTRACTION** of numbers with the *SAME* signs:

4. **RULE: Change the sign of the subtrahend (the second number in a subtraction problem) and proceed as in addition.**

EXAMPLE:

$$\begin{array}{r} +3 \\ -\ \ +2 \\ \hline \end{array} = \begin{array}{r} +3 \\ +\ \ -2 \\ \hline +1 \end{array} \qquad \begin{array}{r} -3 \\ -\ \ -2 \\ \hline \end{array} = \begin{array}{r} -3 \\ ++2 \\ \hline -1 \end{array}$$

E. **MULTIPLICATION:**

5. **RULE: The product of any two numbers having LIKE signs is POSITIVE. The product of any two numbers having UNLIKE signs is NEGATIVE.**

EXAMPLE:

$(+3) \times (-2) = -6$

$(-3) \times (+2) = -6$

$(+3) \times (+2) = +6$

$(-3) \times (-2) = +6$

F. **DIVISION:**

6. **RULE: If the divisor and the dividend have LIKE signs, the sign of the quotient is POSITIVE. If the divisor and dividend have UNLIKE signs, the sign of the quotient is NEGATIVE.**

EXAMPLE:

$\frac{+6}{-2} = -3 \qquad \frac{+6}{+2} = +3$

$\frac{-6}{+2} = -3 \qquad \frac{-6}{-2} = +3$

NATURAL TRIGONOMETRIC FUNCTIONS

ANGLE	SINE	COSINE	TANGENT	COTANGENT	SECANT	COSECANT	
0	.0000	1.0000	.0000		1.0000		90
1	.0175	.9998	.0175	57.2900	1.0002	57.2987	89
2	.0349	.9994	.0349	28.6363	1.0006	28.6537	88
3	.0523	.9986	.0524	19.0811	1.0014	19.1073	87
4	.0698	.9976	.0699	14.3007	1.0024	14.3356	86
5	.0872	.9962	.0875	11.4301	1.0038	11.4737	85
6	.1045	.9945	.1051	9.5144	1.0055	9.5668	84
7	.1219	.9925	.1228	8.1443	1.0075	8.2055	83
8	.1392	.9903	.1405	7.1154	1.0098	7.1853	82
9	.1564	.9877	.1584	6.3138	1.0125	6.3925	81
10	.1736	.9848	.1763	5.6713	1.0154	5.7588	80
11	.1908	.9816	.1944	5.1446	1.0187	5.2408	79
12	.2079	.9781	.2126	4.7046	1.0223	4.8097	78
13	.2250	.9744	.2309	4.3315	1.0263	4.4454	77
14	.2419	.9703	.2493	4.0108	1.0306	4.1336	76
15	.2588	.9659	.2679	3.7321	1.0353	3.8637	75
16	.2756	.9613	.2867	3.4874	1.0403	3.6280	74
17	.2924	.9563	.3057	3.2709	1.0457	3.4203	73
18	.3090	.9511	.3249	3.0777	1.0515	3.2361	72
19	.3256	.9455	.3443	2.9042	1.0576	3.0716	71
20	.3420	.9397	.3640	2.7475	1.0642	2.9238	70
21	.3584	.9336	.3839	2.6051	1.0711	2.7904	69
22	.3746	.9272	.4040	2.4751	1.0785	2.6695	68
23	.3907	.9205	.4245	2.3559	1.0864	2.5593	67
24	.4067	.9135	.4452	2.2460	1.0946	2.4586	66
25	.4226	.9063	.4663	2.1445	1.1034	2.3662	65
26	.4384	.8988	.4877	2.0503	1.1126	2.2812	64
27	.4540	.8910	.5095	1.9626	1.1223	2.2027	63
28	.4695	.8829	.5317	1.8807	1.1326	2.1301	62
29	.4848	.8746	.5543	1.8040	1.1434	2.0627	61
30	.5000	.8660	.5774	1.7321	1.1547	2.0000	60
31	.5150	.8572	.6009	1.6643	1.1666	1.9416	59
32	.5299	.8480	.6249	1.6003	1.1792	1.8871	58
33	.5446	.8387	.6494	1.5399	1.1924	1.8361	57
34	.5592	.8290	.6745	1.4826	1.2062	1.7883	56
35	.5736	.8192	.7002	1.4281	1.2208	1.7434	55
36	.5878	.8090	.7265	1.3764	1.2361	1.7013	54
37	.6018	.7986	.7536	1.3270	1.2521	1.6616	53
38	.6157	.7880	.7813	1.2799	1.2690	1.6243	52
39	.6293	.7771	.8098	1.2349	1.2868	1.5890	51
40	.6428	.7660	.8391	1.1918	1.3054	1.5557	50
41	.6561	.7547	.8693	1.1504	1.3250	1.5243	49
42	.6691	.7431	.9004	1.1106	1.3456	1.4945	48
43	.6820	.7314	.9325	1.0724	1.3673	1.4663	47
44	.6947	.7193	.9657	1.0355	1.3902	1.4396	46
45	.7071	.7071	1.0000	1.0000	1.4142	1.4142	45
	COSINE	SINE	COTANGT.	TANGENT	COSECANT	SECANT	ANGLE

TRIGONOMETRY

TRIGONOMETRY is the mathematics dealing with the relations of sides and angles of triangles.
A **TRIANGLE** is a figure enclosed by three straight sides. The sum of the three angles is 180 degrees. All triangles have six parts: three angles and three sides opposite the angles.
RIGHT TRIANGLES are triangles that have one angle of 90 degrees and two angles of less than 90 degrees.
To help you remember the six trigonometric functions, memorize:

"OH HELL ANOTHER HOUR OF ANDY"

SINE Ø = (*OH*) OPPOSITE SIDE / HYPOTENUSE (*HELL*)

COSINE Ø = (*ANOTHER*) ADJACENT SIDE / HYPOTENUSE (*HOUR*)

TANGENT Ø = (*OF*) OPPOSITE SIDE / ADJACENT SIDE (*ANDY*)

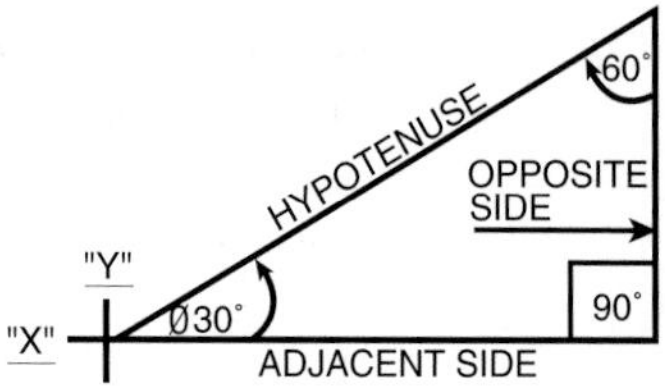

Now, use backwards: **"ANDY OF HOUR ANOTHER HELL OH"**

COTANGENT Ø = (*ANDY*) ADJACENT SIDE / OPPOSITE SIDE (*OF*)

Always place the angle to be solved at the vertex (where "X" and "Y" cross)

SECANT Ø = (*HOUR*) HYPOTENUSE / ADJACENT SIDE (*ANOTHER*)

COSECANT Ø = (*HELL*) HYPOTENUSE / OPPOSITE SIDE (*OH*)

Note:
Ø = Theta = Any Angle

BENDING OFF-SETS WITH TRIGONOMETRY

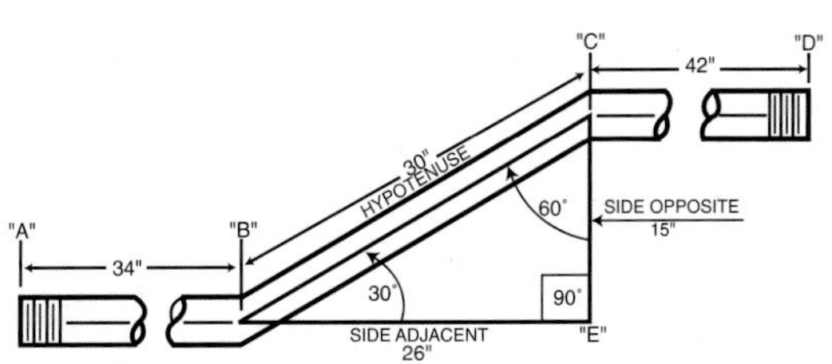

THE COSECANT OF THE ANGLE TIMES THE OFF-SET DESIRED IS EQUAL TO THE DISTANCE BETWEEN THE CENTERS OF THE BENDS.

EXAMPLE:

To make a fifteen inch (15") off-set, using thirty (30) degree bends:

1. Use Trig. Table (page 135) to find the Cosecant of a thirty (30) degree angle. We find it to be two (2).
2. Multiply two (2) times the off-set desired, which is fifteen (15) inches to determine the distance between bend "B" and bend "C". The answer is thirty (30) inches.

To mark the conduit for bending:

1. Measure from end of Conduit "A" thirty-four (34) inches to center of first bend "B", and mark.
2. Measure from mark "B" thirty (30) inches to center of second bend "C" and mark.
3. Measure from mark "C" forty-two (42) inches to "D", and mark. Cut, ream, and thread conduit before bending.

ROLLING OFF-SETS:

To determine how much off-set is needed to make a rolling off-set:

1. Measure vertical required. Use work table (any square will do) and measure from corner this amount and mark.
2. Measure horizontal required. Measure ninety degrees from the vertical line measurement (starting in same corner) and mark.
3. The diagonal distance between these marks will be the amount of off-set required.

Note: Shrink is hypotenuse minus the side adjacent.

CHICAGO-TYPE BENDERS
NINETY DEGREE BENDING

"A" to "C" = STUB-UP
"C" to "D" = TAIL
"C" = BACK OF STUB-UP
"C" = BOTTOM OF CONDUIT

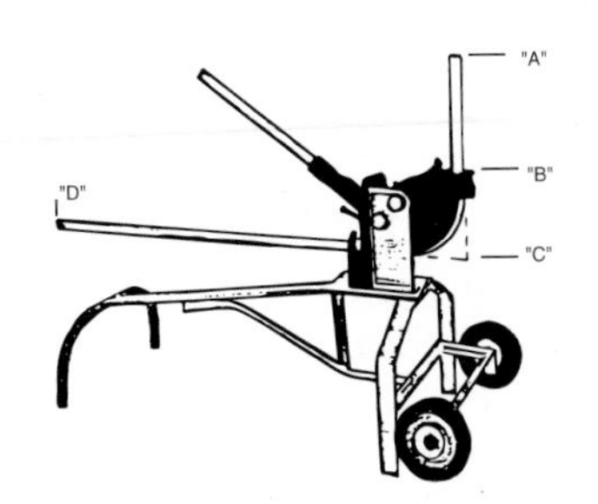

Note:
There are many variations of this type bender, but most manufacturers offer two sizes.
The *small* size shoe takes 1/2", 3/4" and 1" conduit.
The *large* size shoe takes 1¼" and 1½" conduit.

TO DETERMINE THE "TAKE-UP" AND "SHRINK" OF EACH SIZE CONDUIT FOR A PARTICULAR BENDER TO MAKE NINETY DEGREE BENDS:

1. Use a straight piece of scrap conduit.
2. Measure exact length of scrap conduit, "A" to "D".

"D" ORIGINAL MEASUREMENT "A"

3. Place conduit in bender. Mark at edge of shoe, "B".
4. Level conduit. Bend ninety, and count number of pumps. Be sure to keep notes on each size conduit used.
5. After bending ninety:
   A. Distance between "B" and "C" is the TAKE-UP.
   B. Original measurement of the scrap piece of conduit subtracted from (distance "A" to "C" plus distance "C" to "D") is the SHRINK.

Note: Both time and energy will be saved if conduit can be cut, reamed and threaded before bending.

The same method can be used on hydraulic benders.

CHICAGO-TYPE BENDERS
OFF-SETS

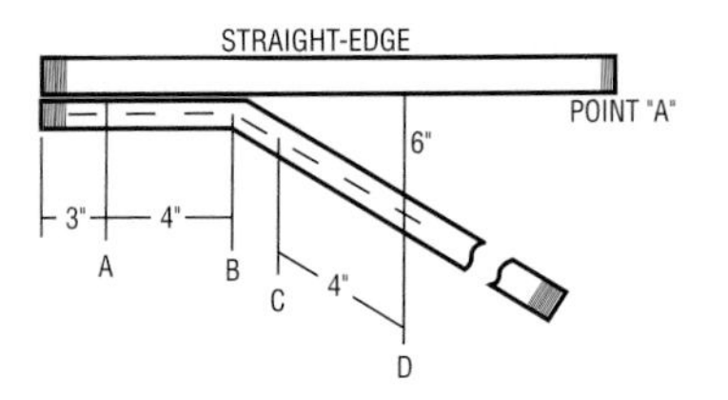

CHICAGO TYPE BENDER

EXAMPLE: To bend a 6" off-set:

1. Make a mark 3" from conduit end. Place conduit in bender with mark at outside edge of jaw.
2. Make three full pumps, making sure handle goes all the way down to the stop.
3. Remove conduit from bender and place alongside straight-edge.
4. Measure 6" from straight-edge to center of conduit. Mark point "D". Use square for accuracy.
5. Mark center of conduit from both directions through bend as shown by broken line. Where lines intersect is point "B".
6. Measure from "A" to "B" to determine distance from "D" to "C". Mark "C" and place conduit in bender with mark at outside edge of jaw, and with the kick pointing down. Use a level to prevent dogging conduit.
7. Make three full pumps, making sure handle goes all the way down to the stop.

Note: 1. There are several methods of bending rigid conduit with a Chicago Type Bender, and any method that gets the job done in a minimal amount of time with craftsmanship is acceptable.

2. Whatever method is used, quality will improve with experience.

MULTI-SHOT NINETY DEGREE CONDUIT BENDING

PROBLEM:

A. To measure, thread, cut and ream conduit before bending.

B. To accurately bend conduit to the desired height of the stub-up (H), and to the desired length of the tail (L).

GIVEN:

A. Size of conduit = 2"

B. Space between conduit (center to center) = 6"

C. Height of stub-up = 36"

D. Length of tail = 48"

SOLUTION:

A. TO DETERMINE RADIUS (R):

Conduit #1 (inside conduit) will use the minimum radius unless otherwise specified. The minimum radius is eight times the size of the conduit. (see page 142)

RADIUS OF CONDUIT #1 = 8 x 2" + 1.25" = 17.25"

RADIUS OF CONDUIT #2 = RADIUS #1 + 6" = 23.25"

RADIUS OF CONDUIT #3 = RADIUS #2 + 6" = 29.25"

B. TO DETERMINE DEVELOPED LENGTH (DL): RADIUS X 1.57 = DL

DL OF CONDUIT #1 = R x 1.57 = 17.25" x 1.57 = 27"

DL OF CONDUIT #2 = R x 1.57 = 23.25" x 1.57 = 36.5"

DL OF CONDUIT #3 = R x 1.57 = 29.25" x 1.57 = 46"

C. TO DETERMINE LENGTH OF NIPPLE:

LENGTH OF NIPPLE, CONDUIT #1 = L + H + DL - 2R
= 48" + 36" + 27" - 34.5"
= 76.5"

LENGTH OF NIPPLE, CONDUIT #2 = L + H + DL - 2R
= 54" + 42" + 36.5" - 46.5"
= 86"

LENGTH OF NIPPLE, CONDUIT #3 = L + H + DL - 2R
= 60" + 48" + 46" - 58.5"
= 95.5"

Note: 1. For 90 degree bends, SHRINK = 2R - DL
2. For off-set bends, SHRINK = HYPOTENUSE - SIDE ADJACENT

MULTI-SHOT NINETY DEGREE CONDUIT BENDING

LAYOUT AND BENDING:

A. To locate point "B", measure from point "A", the length of the stub-up minus the radius. On all three conduit, point "B" will be 18.75" from point "A". (see page 142).

B. To locate point "C", measure from point "D", the length minus the radius, (see page 142). On all three conduit, point "C" will be 30.75" from point "D". (see page 142).

C. Divide the developed length (point "B" to point "C") into equal spaces. Spaces should not be more than 1.75" to prevent wrinkling of the conduit. On Conduit #1, seventeen spaces of 1.5882" each would give us eighteen shots of 5 degrees each. Remember there is always one less space than shot. When determining the number of shots, choose a number that will divide into ninety an even number of times.

D. If an elastic numbered tape is not available, try the method illustrated.

A to B = Conduit #1
Developed Length = 27"

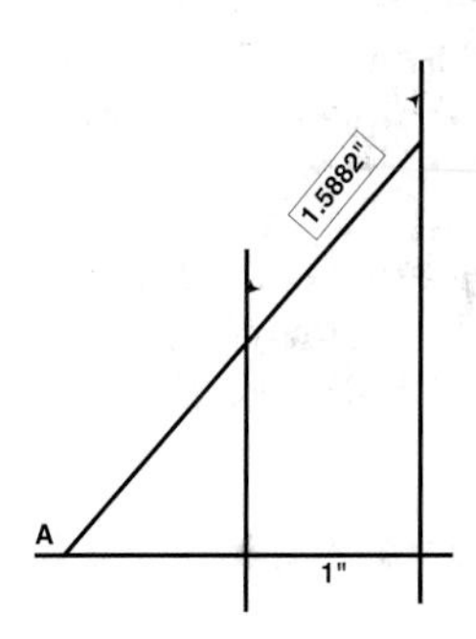

A to C = 17 1" spaces
A to B = 17 1.5882" spaces
C = table or plywood corner

Measure from Point "C" (table corner) 17 inches along table edge to Point "A" and mark. Place end of rule at Point "A". Point "B" will be located where 27" mark meets table edge B-C. Mark on board, then transfer to conduit.

MULTI-SHOT NINETY DEGREE CONDUIT BENDING

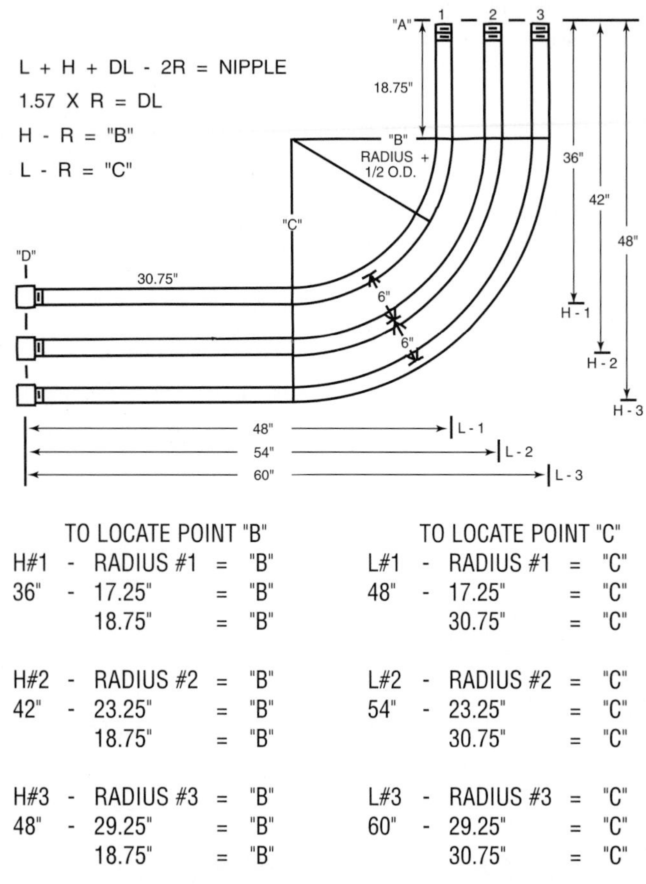

TO LOCATE POINT "B"

H#1 - RADIUS #1 = "B"
36" - 17.25" = "B"
18.75" = "B"

H#2 - RADIUS #2 = "B"
42" - 23.25" = "B"
18.75" = "B"

H#3 - RADIUS #3 = "B"
48" - 29.25" = "B"
18.75" = "B"

TO LOCATE POINT "C"

L#1 - RADIUS #1 = "C"
48" - 17.25" = "C"
30.75" = "C"

L#2 - RADIUS #2 = "C"
54" - 23.25" = "C"
30.75" = "C"

L#3 - RADIUS #3 = "C"
60" - 29.25" = "C"
30.75" = "C"

Points "B" and "C" are the same distance from the end on all three conduits.

OFFSET BENDS - EMT - USING HAND BENDER

An offset bend is used to change the level, or plane, of the conduit. This is usually necessitated by the presence of an obstruction in the original conduit path.

Step One:

Determine the offset depth. (X)

Step Two:

Multiply the offset depth "x" the multiplier for the degree of bend used to determine the distance between bends.

ANGLE		MULTIPLIER
10° x 10°	=	6
22½° x 22½°	=	2.6
30° x 30°	=	2
45° x 45°	=	1.4
60° x 60°	=	1.2

Example: If the offset depth required (X) is 6", and you intend to use 30° bends, the distance between bends is 6" x 2 = 12".

Step Three:

Mark at the appropriate points, align the arrow on the bender with the first mark, and bend to desired degree by aligning EMT with chosen degree line on bender.

Step Four:

Slide down the EMT, align the arrow with the second mark, and bend to the same degree line. Be sure to note the orientation of the bender head. Check alignment.

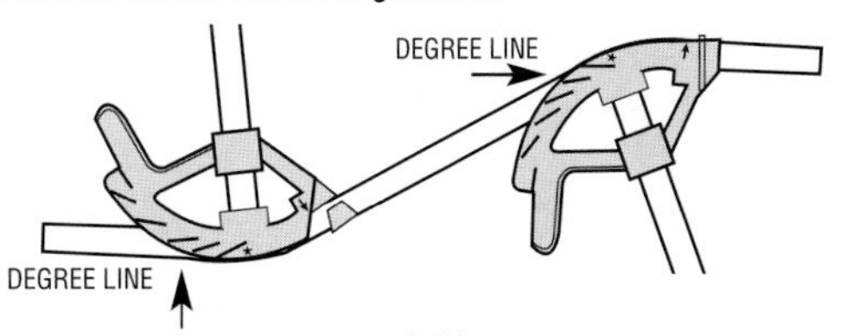

90° BENDS - EMT - USING HAND BENDER

The stub is the most common bend.

Step One:

Determine the height of the stub-up required, and mark on EMT.

Step Two:

Find the "Deduct" or "Take-up" amount from the Take-Up Chart. Subtract the take-up amount from the stub height, and mark the EMT that distance from the end.

Step Three:

Align the arrow on bender with the last mark made on the EMT, and bend to the 90° mark on the bender.

DESCRIPTION		TAKE-UP
1/2" EMT	=	5"
3/4" EMT	=	6"
1" EMT	=	8"
1-1/4" EMT	=	11"

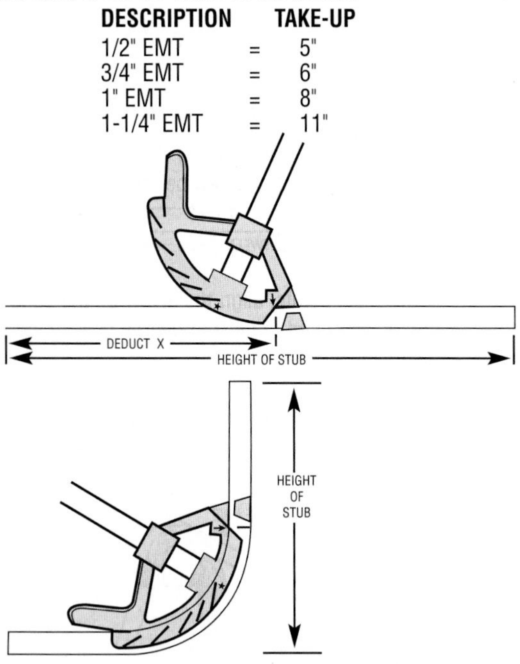

BACK-TO-BACK BENDS - EMT- USING HAND BENDER

A back-to-back bend results in a "U" shape in a length of conduit. It's used for a conduit which runs along the floor or ceiling which turns up or down a wall.

Step One:

After the first 90° bend is made, determine the back-to-back length and mark on EMT.

Step Two:

Align this back-to-back mark with the star mark on the bender, and bend to 90°.

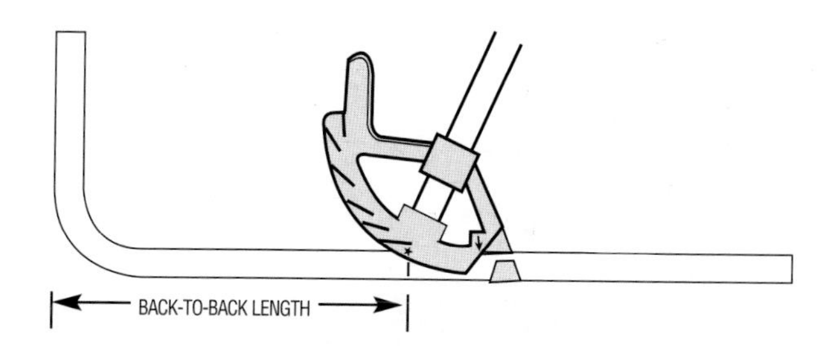

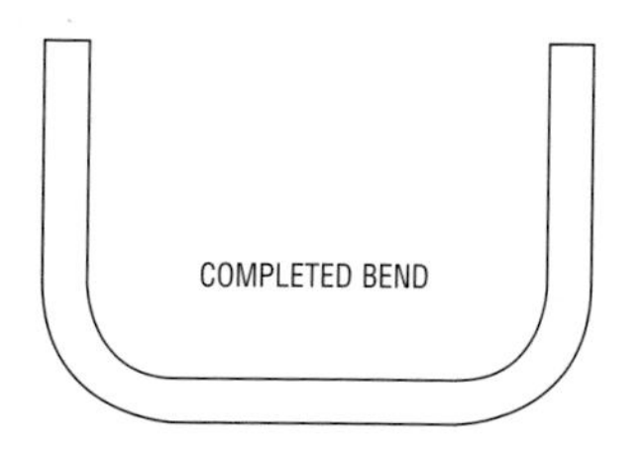

3-POINT SADDLE BENDS - EMT- USING HAND BENDER

The 3-point saddle bend is used when encountering an obstacle (usually another pipe)

Step One:

Measure the height of the obstruction.
Mark the center point on EMT.

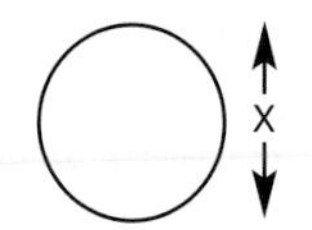

Step Two:

Multiply the height of the obstruction by 2.5 and mark this distance on each side of the center mark.

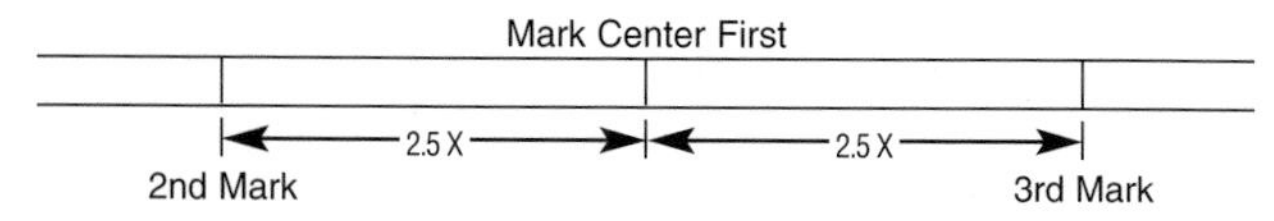

Step Three:

Place the center mark on the saddle mark or notch. Bend to 45°.

Step Four:

Bend the second mark to 22-1/2° angle at arrow.

Step Five:

Bend the third mark to 22-1/2° angle at arrow. Be aware of the orientation of the EMT on all bends. Check alignment.

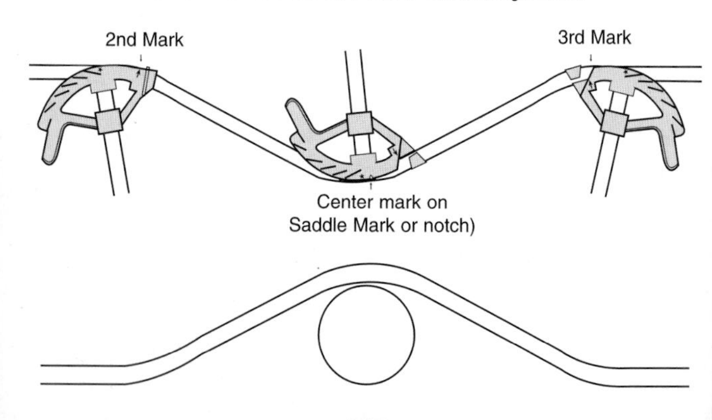

PULLEY CALCULATIONS

The most common configuration consists of a motor with a pulley attached to its shaft, connected by a belt to a second pulley. The motor pulley is referred to as the **Driving Pulley**. The second pulley is called the **Driven Pulley**. The speed at which the Driven Pulley turns is determined by the speed at which the Driving Pulley turns as well as the diameters of both pulleys. The following formulas may be used to determine the relationships between the motor, pulley diameters and pulley speeds.

D = **Diameter of Driving Pulley**
d^1 = **Diameter of Driven Pulley**
S = **Speed of Driving Pulley** (revolutions per minute)
s^1 = **Speed of Driven Pulley** (revolutions per minute)

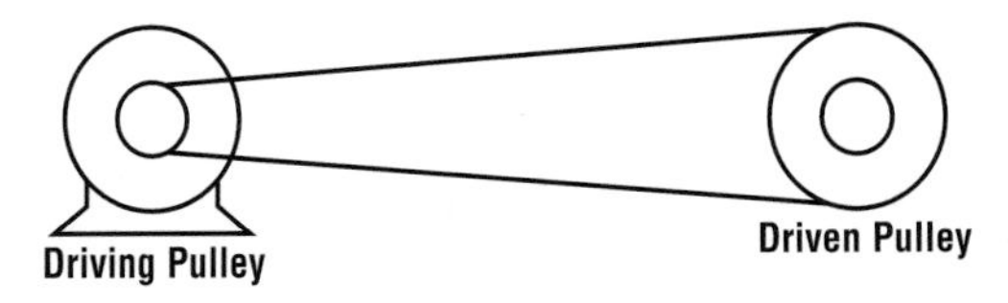

- *To determine the speed of the Driven Pulley (Driven RPM):*

$$s^1 = \frac{D \times S}{d^1} \quad \text{or} \quad \text{Driven RPM} = \frac{\text{Driving Pulley Dia.} \times \text{Driving RPM}}{\text{Driven Pulley Dia.}}$$

- *To determine the speed of the Driving Pulley (Driving RPM):*

$$S = \frac{d^1 \times s^1}{D} \quad \text{or} \quad \text{Driving RPM} = \frac{\text{Driven Pulley Dia.} \times \text{Driven RPM}}{\text{Driving Pulley Dia.}}$$

- *To determine the diameter of the Driven Pulley (Driven Dia.):*

$$d^1 = \frac{D \times S}{s^1} \quad \text{or} \quad \text{Driven Dia.} = \frac{\text{Driving Pulley Dia.} \times \text{Driving RPM}}{\text{Driven RPM}}$$

- *To determine the diameter of the Driving Pulley (Driving Dia.):*

$$D = \frac{d^1 \times s^1}{S} \quad \text{or} \quad \text{Driving Dia.} = \frac{\text{Driven Pulley Dia.} \times \text{Driven RPM}}{\text{Driving RPM}}$$

USEFUL KNOTS

BOWLINE

RUNNING BOWLINE

BOWLINE ON THE BIGHT

CLOVE HITCH

SHEEPSHANK

ROLLING HITCH

SINGLE BLACKWALL HITCH

CATSPAW

DOUBLE BLACKWALL HITCH

SQUARE KNOT

TIMBER HITCH WITH HALF HITCH

SINGLE SHEET BEND

HAND SIGNALS

STOP

DOG EVERYTHING

EMERGENCY STOP

TRAVEL

TRAVEL BOTH TRACKS (CRAWLER CRANES ONLY)

TRAVEL ONE TRACK (CRAWLERS)

RETRACT BOOM

EXTEND BOOM

SWING BOOM

HAND SIGNALS

RAISE LOAD

LOWER LOAD

MAIN HOIST

MOVE SLOWLY

RAISE BOOM AND LOWER LOAD (FLEX FINGERS)

LOWER BOOM AND RAISE LOAD (FLEX FINGERS)

USE WHIP LINE

BOOM UP

BOOM DOWN

FIRST AID

The American Red Cross recommends certification in a CPR training course annually and certification in a first aid course every three years. (Also refer to author's and publisher's disclaimer on inside front cover)

GENERAL DIRECTIONS FOR FIRST AID:

While help is being summoned, do the following:
1) CHECK - Check the scene and the injured person.
2) CALL - Call 9-1-1 or the local emergency number.
3) CARE - Care for the injured person.

URGENT CARE:

BLEEDING

First Aid:

1) Direct Pressure and Elevation:
   * Place dressing and apply pressure directly over the wound, then elevate above the level of the heart, unless there is evidence of a fracture.
2) Apply pressure bandage.
   * Wrap bandage snugly over the dressing.
3) Pressure Points
   * If bleeding doesn't stop after direct pressure, elevation and the pressure bandage, compress the pressure point.
   * Arm: Use the brachial artery - pushing the artery against the upper arm bone.
   * Leg: Apply pressure on femoral artery, pushing it against the pelvic bone.
4) Nosebleed:
   To control a nosebleed, have the victim lean forward and pinch the nostrils together until bleeding stops.

POISONING

Signals: Vomiting, heavy labored breathing, sudden onset of pain or illness, burns or odor around the lips or mouth, unusual behavior.

(Poisoning, continued)

First Aid:

*** If you think someone has been poisoned, call your poison control center or local emergency number and follow their directions.**

* Try to identify the poison - be prepared to inform poison center of the type of poison, when incident occurred, victim's age, symptoms, and how much poison may have been ingested, inhaled, absorbed, or injected.

If unconscious or nauseous:

1) Position victim on side and monitor vital signs (i.e. pulse and breathing).
2) Call Poison Control and identify the poison.
3) DO NOT give anything by mouth.

SHOCK

Signals: Cool, moist, pale, bluish skin, weak rapid pulse (over 100), nausea, rate of breathing increased, apathetic.

First Aid:

Caring for shock involves the following simple steps:

1) Have the victim lie down in the most comfortable position to minimize any pain. Pain can intensify the body's stress and accelerate the progression of shock. Help the victim to rest comfortably.
2) Control any external bleeding
3) Help the victim maintain normal body temperature. If the victim is cool, try to cover to avoid chill.
4) Try to reassure the victim.
5) Elevate the legs about 12" unless you suspect head, neck back injuries or possible broken bones involving the hips or legs. If unsure of the victim's condition, leave the victim lying flat.
6) Do not give the victim anything to eat or drink.
7) Call 911 or your local emergency number immediately.

Shock **cannot** be managed effectively by first aid alone. A victim of shock requires advanced medical care as soon as possible.

BURNS

Signals: Small or large thin (surface) burns:
redness, pain and swelling.
Deep burns: blisters, deep tissue destruction, charred appearance.

First Aid:

1) Stop the burning - put out flames or remove the victim from the source of the burn.
2) Cool all burns (except electrical) - run or pour cool water on burn. Immerse if possible. Cool until pain is reduced.
3) Cover the burn - Use dry, sterile dressing and bandage.
4) Keep victim comfortable as possible, **not** chilled or over heated.

Chemical burn - must be flushed with large amounts of water until EMS arrives.
Electrical burn - make sure power is turned off before touching the victim. Care for life threatening injury.

ELECTRICAL SHOCK

Signals: Unconsciousness, absence of breathing & pulse.
First Aid:

1) TURN OFF THE POWER SOURCE - Call EMS.
(DO NOT approach the victim until power has been turned off.)
2) DO NOT move a victim of electrical injury unless there is immediate danger.
3) Administer rescue breathing or CPR if necessary.
4) Treat for shock.
5) Check for other injuries and monitor victim until Medical help arrives.

FROSTBITE

Signals: Flushed, white or gray skin. Pain. The nose, cheeks, ears, fingers, and toes are most likely to be affected. Pain may be felt early and then subside. Blisters may appear later.

(Frostbite, continued)

First Aid:

1) Cover the frozen part. Loosen restrictive clothing or boots.
2) Bring victim indoors ASAP.
3) Give the victim a warm drink. (DO NOT give alcoholic beverages, tea, or coffee)
4) Immerse frozen part in warm water (100° - 105°), or wrap in a sheet and warm blankets. DO NOT rewarm if there is a possibility of refreezing.
5) Remove from water and discontinue warming once part becomes flushed.
6) Elevate the injured area and protect from further injury.
7) DO NOT rub the frozen part. DO NOT break the blisters. DO NOT use extreme or dry heat to rewarm the part.
8) If fingers or toes are involved, place dry, sterile gauze between them when bandaging.

HYPOTHERMIA

Signals: Lowered body core temperature. Persistent shivering, lips may be blue, slow slurred speech. memory lapses. Most cases occur when air temperature ranges from 30° - 50° or water temperature is below 70°F.

First Aid:

1) Move victim to shelter and remove wet clothing if necessary.
2) Rewarm victim with blankets or body-to-body contact in sleeping bag.
3) If victim is conscious and able to swallow, give warm liquids.
4) Keep victim warm and quiet.
5) DO NOT give alcoholic beverages, or beverages containing caffeine.
6) Constantly monitor victim and give Rescue Breathing and CPR if necessary.

HEAT EXHAUSTION / HEAT STROKE

Signals: *Heat Exhaustion:* Pale, clammy skin, profuse perspiration, weakness, nausea, headache.

Heat Stroke: Hot dry red skin, no perspiration, rapid & weak pulse. High body temperature (105°+).

This is an immediate life threatening emergency; Call 911.

First Aid:

1) Get the victim out of the heat.
2) Loosen tight clothing or restrictive clothing.
3) Remove perspiration soaked clothing.
4) Apply cool, wet cloths to the skin.
5) Fan the victim.
6) If victim is conscious, give cool water to drink.
7) Call for an ambulance if victim refuses water, vomits, or starts to lose consciousness.

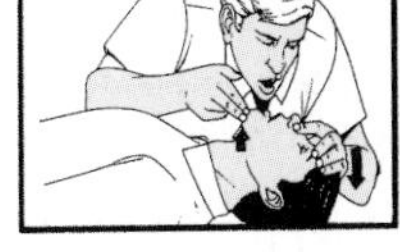

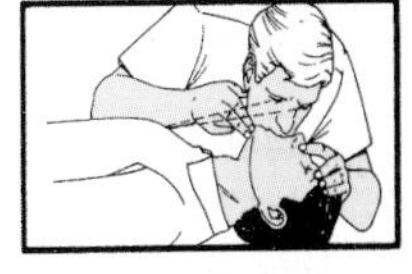

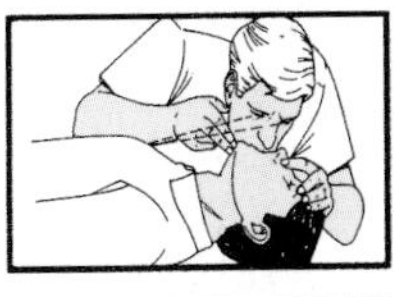

RESCUE BREATHING

1. **CHECK THE VICTIM**

Tap and shout "Are you okay?", to see if the person responds.

If no response...

2. **CALL EMS.**

3. **CARE FOR THE VICTIM.**

Step 1: Look, listen and feel for breathing for about 5 seconds.

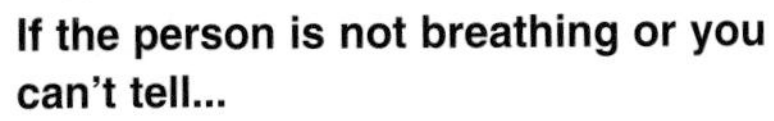

If the person is not breathing or you can't tell...

Step 2: Position victim on back, while supporting head and neck.

Step 3: Tilt head back and lift chin.

Step 4: Look, listen, and feel for breathing for about 5 seconds.

If not breathing...

Step 5: Give two slow gentle breaths.

Step 6: Check signs of circulation (pulse) for 5 to 10 seconds.

Step 7: Check for severe bleeding.

4. **GIVE RESCUE BREATHING**

If pulse is present but person is still not breathing...

Step 1: Give one slow breath about every five seconds. Do this for about one minute (12 breaths).

Step 2: Recheck signs of circulation (pulse) and breathing about every minute.

Continue rescue breathing as long as pulse is present but person is not breathing.

5. BEGIN CARDIOPULMONARY RESUSCITATION (CPR) If there is no sign of circulation (pulse) and no breathing.

FIRST AID FOR CHOKING

1. Check the victim.

When an adult is choking:

Step 1: Ask, **"Are you choking?"** If victim cannot cough, speak, or breathe, is coughing weakly or is making high-pitched noises...

Step 2: Shout **"HELP!"**.

Step 3: **Phone EMS** for help - Send someone to call for an ambulance.

Step 4: **Do Abdominal Thrusts.**

A. Wrap your arms around victim's waist. Make a fist. Place thumbside of fist against middle of abdomen just above the navel. Grasp fist with other hand.

B. Give quick, upward thrusts.

Repeat until object is coughed up or person becomes unconscious.

Be Prepared To Save A Life

Be prepared for an accident with an American Red Cross Pillow First Aid Kit.

The American Red Cross Pillow First Aid Kit makes it easy to treat almost any emergency quickly and correctly.

The kit contains sealed packets and printed step-by-step directions. It's ideal for storing inside your auto and also serves as a pillow.

Count on the Red Cross to back you up with a complete family first aid kit for your car, home, truck, boat or workplace. For more information contact your local Red Cross chapter. In Houston, Texas and surrounding counties, call (713) 526-8300.

Together, we can save a life

Now you'll be prepared, when it's up to you.

The American Red Cross recommends CPR certification annually and first aid certification every three years.

www.houstonredcross.org

This advertisement donated by Burleson Distributing Corporation

2002 CALENDAR

JANUARY

S	M	T	W	T	F	S
		1	2	3	4	5
6	7	8	9	10	11	12
13	14	15	16	17	18	19
20	21	22	23	24	25	26
27	28	29	30	31		

FEBRUARY

S	M	T	W	T	F	S
					1	2
3	4	5	6	7	8	9
10	11	12	13	14	15	16
17	18	19	20	21	22	23
24	25	26	27	28		

MARCH

S	M	T	W	T	F	S
					1	2
3	4	5	6	7	8	9
10	11	12	13	14	15	16
17	18	19	20	21	22	23
24	25	26	27	28	29	30
31						

APRIL

S	M	T	W	T	F	S
	1	2	3	4	5	6
7	8	9	10	11	12	13
14	15	16	17	18	19	20
21	22	23	24	25	26	27
28	29	30				

MAY

S	M	T	W	T	F	S
			1	2	3	4
5	6	7	8	9	10	11
12	13	14	15	16	17	18
19	20	21	22	23	24	25
26	27	28	29	30	31	

JUNE

S	M	T	W	T	F	S
						1
2	3	4	5	6	7	8
9	10	11	12	13	14	15
16	17	18	19	20	21	22
23	24	25	26	27	28	29
30						

JULY

S	M	T	W	T	F	S
	1	2	3	4	5	6
7	8	9	10	11	12	13
14	15	16	17	18	19	20
21	22	23	24	25	26	27
28	29	30	31			

AUGUST

S	M	T	W	T	F	S
				1	2	3
4	5	6	7	8	9	10
11	12	13	14	15	16	17
18	19	20	21	22	23	24
25	26	27	28	29	30	31

SEPTEMBER

S	M	T	W	T	F	S
1	2	3	4	5	6	7
8	9	10	11	12	13	14
15	16	17	18	19	20	21
22	23	24	25	26	27	28
29	30					

OCTOBER

S	M	T	W	T	F	S
		1	2	3	4	5
6	7	8	9	10	11	12
13	14	15	16	17	18	19
20	21	22	23	24	25	26
27	28	29	30	31		

NOVEMBER

S	M	T	W	T	F	S
					1	2
3	4	5	6	7	8	9
10	11	12	13	14	15	16
17	18	19	20	21	22	23
24	25	26	27	28	29	30

DECEMBER

S	M	T	W	T	F	S
1	2	3	4	5	6	7
8	9	10	11	12	13	14
15	16	17	18	19	20	21
22	23	24	25	26	27	28
29	30	31				

2003 CALENDAR

JANUARY

S	M	T	W	T	F	S
			1	2	3	4
5	6	7	8	9	10	11
12	13	14	15	16	17	18
19	20	21	22	23	24	25
26	27	28	29	30	31	

FEBRUARY

S	M	T	W	T	F	S
						1
2	3	4	5	6	7	8
9	10	11	12	13	14	15
16	17	18	19	20	21	22
23	24	25	26	27	28	

MARCH

S	M	T	W	T	F	S
						1
2	3	4	5	6	7	8
9	10	11	12	13	14	15
16	17	18	19	20	21	22
23	24	25	26	27	28	29
30	31					

APRIL

S	M	T	W	T	F	S
		1	2	3	4	5
6	7	8	9	10	11	12
13	14	15	16	17	18	19
20	21	22	23	24	25	26
27	28	29	30			

MAY

S	M	T	W	T	F	S
				1	2	3
4	5	6	7	8	9	10
11	12	13	14	15	16	17
18	19	20	21	22	23	24
25	26	27	28	29	30	31

JUNE

S	M	T	W	T	F	S
1	2	3	4	5	6	7
8	9	10	11	12	13	14
15	16	17	18	19	20	21
22	23	24	25	26	27	28
29	30					

JULY

S	M	T	W	T	F	S
		1	2	3	4	5
6	7	8	9	10	11	12
13	14	15	16	17	18	19
20	21	22	23	24	25	26
27	28	29	30	31		

AUGUST

S	M	T	W	T	F	S
					1	2
3	4	5	6	7	8	9
10	11	12	13	14	15	16
17	18	19	20	21	22	23
24	25	26	27	28	29	30
31						

SEPTEMBER

S	M	T	W	T	F	S
	1	2	3	4	5	6
7	8	9	10	11	12	13
14	15	16	17	18	19	20
21	22	23	24	25	26	27
28	29	30				

OCTOBER

S	M	T	W	T	F	S
			1	2	3	4
5	6	7	8	9	10	11
12	13	14	15	16	17	18
19	20	21	22	23	24	25
26	27	28	29	30	31	

NOVEMBER

S	M	T	W	T	F	S
						1
2	3	4	5	6	7	8
9	10	11	12	13	14	15
16	17	18	19	20	21	22
23	24	25	26	27	28	29
30						

DECEMBER

S	M	T	W	T	F	S
	1	2	3	4	5	6
7	8	9	10	11	12	13
14	15	16	17	18	19	20
21	22	23	24	25	26	27
28	29	30	31			

2004 CALENDAR

JANUARY

S	M	T	W	T	F	S
				1	2	3
4	5	6	7	8	9	10
11	12	13	14	15	16	17
18	19	20	21	22	23	24
25	26	27	28	29	30	31

FEBRUARY

S	M	T	W	T	F	S
1	2	3	4	5	6	7
8	9	10	11	12	13	14
15	16	17	18	19	20	21
22	23	24	25	26	27	28
29						

MARCH

S	M	T	W	T	F	S
	1	2	3	4	5	6
7	8	9	10	11	12	13
14	15	16	17	18	19	20
21	22	23	24	25	26	27
28	29	30	31			

APRIL

S	M	T	W	T	F	S
				1	2	3
4	5	6	7	8	9	10
11	12	13	14	15	16	17
18	19	20	21	22	23	24
25	26	27	28	29	30	

MAY

S	M	T	W	T	F	S
						1
2	3	4	5	6	7	8
9	10	11	12	13	14	15
16	17	18	19	20	21	22
23	24	25	26	27	28	29
30	31					

JUNE

S	M	T	W	T	F	S
		1	2	3	4	5
6	7	8	9	10	11	12
13	14	15	16	17	18	19
20	21	22	23	24	25	26
27	28	29	30			

JULY

S	M	T	W	T	F	S
				1	2	3
4	5	6	7	8	9	10
11	12	13	14	15	16	17
18	19	20	21	22	23	24
25	26	27	28	29	30	31

AUGUST

S	M	T	W	T	F	S
1	2	3	4	5	6	7
8	9	10	11	12	13	14
15	16	17	18	19	20	21
22	23	24	25	26	27	28
29	30	31				

SEPTEMBER

S	M	T	W	T	F	S
			1	2	3	4
5	6	7	8	9	10	11
12	13	14	15	16	17	18
19	20	21	22	23	24	25
26	27	28	29	30		

OCTOBER

S	M	T	W	T	F	S
					1	2
3	4	5	6	7	8	9
10	11	12	13	14	15	16
17	18	19	20	21	22	23
24	25	26	27	28	29	30
31						

NOVEMBER

S	M	T	W	T	F	S
	1	2	3	4	5	6
7	8	9	10	11	12	13
14	15	16	17	18	19	20
21	22	23	24	25	26	27
28	29	30				

DECEMBER

S	M	T	W	T	F	S
			1	2	3	4
5	6	7	8	9	10	11
12	13	14	15	16	17	18
19	20	21	22	23	24	25
26	27	28	29	30	31	

GENERAL ALPHABETICAL INDEX

* Indicates Tables

- Detach along perforations -

UGLY'S 2002 REGISTRATION CARD

Please enter my name into your UGLY'S database, and contact me with information on new UGLY'S products/updates as they become available.

3

Name: ______________________________

Company Name: (if purchased in name of company) ______________________________

Mailing Address: ______________________________

City: ____________________ State: ___ Zip: __________

Phone: (____) __________ ext.: ______ Fax: (____) __________

E-mail address: ______________________________

Please check the box that best describes your job function:

❑ Electrical Contractor ❑ Engineering/Design ❑ Plant/Facilities ❑ Maintenance/Repair

❑ Instructor ❑ Student ❑ Other ____________________

Please check the box that best describes how you acquired this UGLY'S product:

❑ Electrical Supply Co. ❑ Bookstore ❑ Burleson Distributing ❑ Gift

❑ Other ____________________

- Detach along perforations -

To receive information about new UGLY'S products in the future, complete and return this registration card.

Place
Stamp
Here

BURLESON DISTRIBUTING CORPORATION
3501 OAK FOREST DRIVE
HOUSTON, TX 77018-6121